On Course Mapping Instruction

with Lesson Plans for Universal Access

Contents

About This Booklet ii

Program Resources vi

CHAPTER 1 The Nature of Earth Science 2

CHAPTER 2 Tools of Earth Science 12

CHAPTER 3 Earth's Systems and Cycles 24

CHAPTER 4 Material Resources 36

CHAPTER 5 Energy Resources 46

CHAPTER 6 Plate Tectonics 54

CHAPTER 7 Earthquakes 66

CHAPTER 8 Volcanoes 76

CHAPTER 9 Weathering and Soil Formation 86

CHAPTER 10 Agents of Erosion and Deposition 98

CHAPTER 11 Rivers and Groundwater 110

CHAPTER 12 Exploring the Oceans 120

CHAPTER 13 The Movement of Ocean Water 132

CHAPTER 14 The Atmosphere 142

CHAPTER 15 Weather and Climate 154

CHAPTER 16 Interactions of Living Things 166

CHAPTER 17 Biomes and Ecosystems 176

Unpacking the Standards 188

About This Booklet

The charts contained in this booklet will help you tailor your science instruction to a wide range of students. Three types of charts are provided for each chapter: one for Getting Started, one for each section, and one for Wrapping Up. The components of these charts and the Universal Access categories are described below.

The **Pacing** item estimates the amount of time needed to cover a chapter or part of a chapter.

This row, which is from the Standards Course of Study pages in the Teacher's Edition, lists the activities, labs, and resources that provide complete instruction on the California Science Standards covered.

These rows identify activities and resources that are appropriate for students who have the learning needs identified in the legend below.

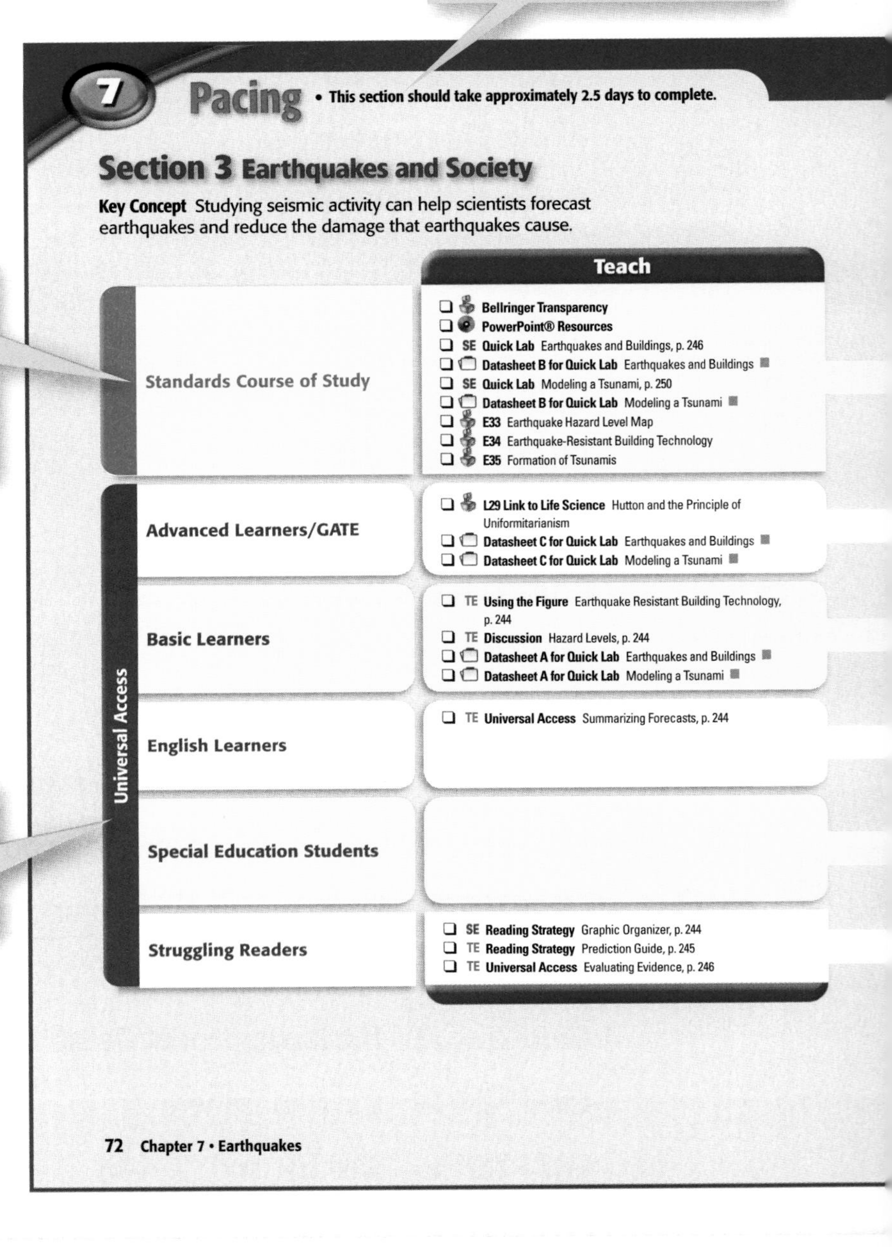

7 **Pacing** • This section should take approximately 2.5 days to complete.

Section 3 Earthquakes and Society

Key Concept Studying seismic activity can help scientists forecast earthquakes and reduce the damage that earthquakes cause.

	Teach
Standards Course of Study	❑ **Bellringer Transparency** ❑ **PowerPoint® Resources** ❑ SE **Quick Lab** Earthquakes and Buildings, p. 246 ❑ **Datasheet B for Quick Lab** Earthquakes and Buildings ■ ❑ SE **Quick Lab** Modeling a Tsunami, p. 250 ❑ **Datasheet B for Quick Lab** Modeling a Tsunami ■ ❑ **E33** Earthquake Hazard Level Map ❑ **E34** Earthquake-Resistant Building Technology ❑ **E35** Formation of Tsunamis
Advanced Learners/GATE	❑ **L29 Link to Life Science** Hutton and the Principle of Uniformitarianism ❑ **Datasheet C for Quick Lab** Earthquakes and Buildings ■ ❑ **Datasheet C for Quick Lab** Modeling a Tsunami ■
Basic Learners	❑ TE **Using the Figure** Earthquake Resistant Building Technology, p. 244 ❑ TE **Discussion** Hazard Levels, p. 244 ❑ **Datasheet A for Quick Lab** Earthquakes and Buildings ■ ❑ **Datasheet A for Quick Lab** Modeling a Tsunami ■
English Learners	❑ TE **Universal Access** Summarizing Forecasts, p. 244
Special Education Students	
Struggling Readers	❑ SE **Reading Strategy** Graphic Organizer, p. 244 ❑ TE **Reading Strategy** Prediction Guide, p. 245 ❑ TE **Universal Access** Evaluating Evidence, p. 246

Universal Access

72 Chapter 7 • Earthquakes

Categories for Universal Access

Basic Learners Students at this level can manage only core assignments. These students have difficulty understanding abstract or complex concepts and may be limited in their ability to learn. Struggling readers and students who have certain learning disabilities also fall into this category.

Advanced Learners/GATE Students at this level are able to master honors or pre-AP coursework. These students are capable of solving hypothetical problems and applying deductive logic. They can readily master the challenge of critical-thinking and synthesis activities.

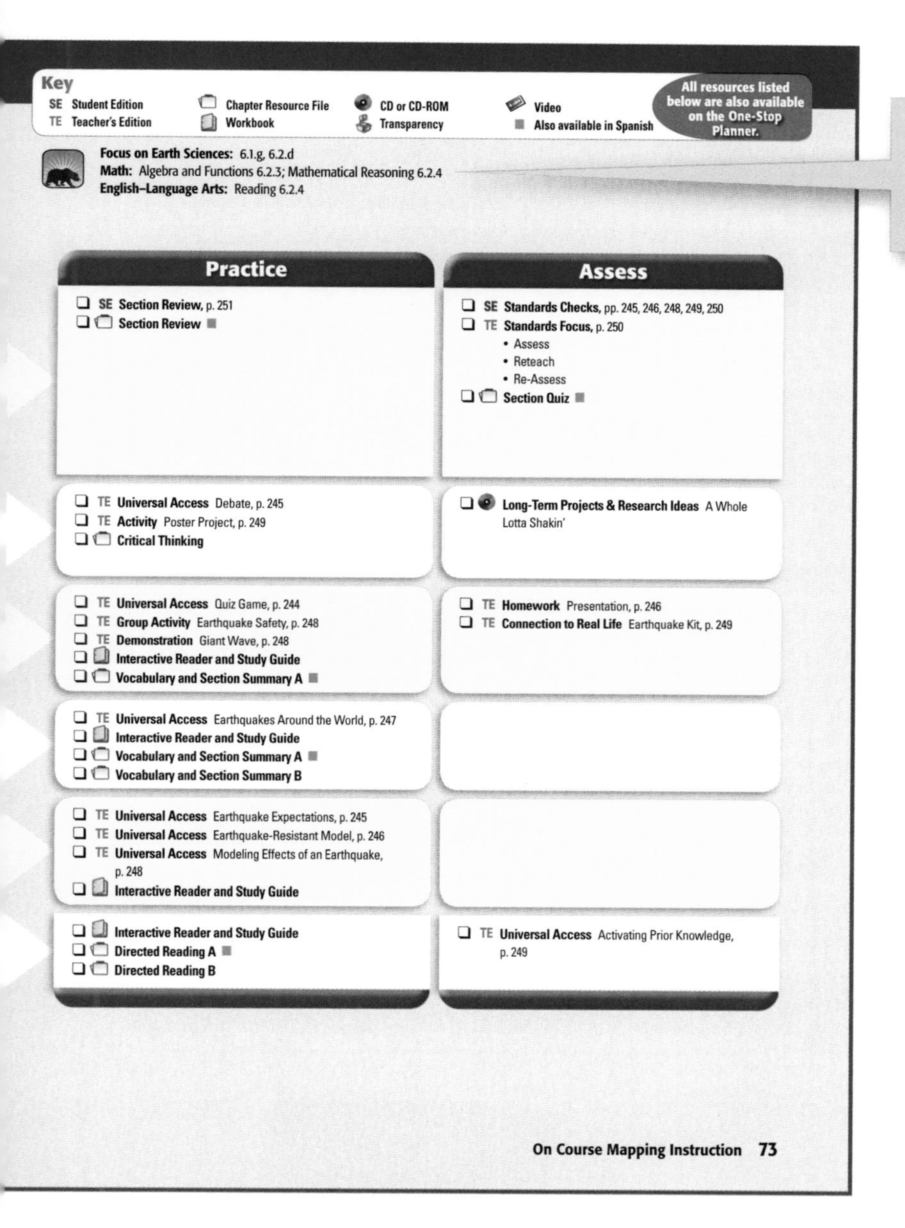

Key
SE Student Edition
TE Teacher's Edition
Chapter Resource File
Workbook
CD or CD-ROM
Transparency
Video
■ Also available in Spanish
All resources listed below are also available on the One-Stop Planner.

Focus on Earth Sciences: 6.1.g, 6.2.d
Math: Algebra and Functions 6.2.3; Mathematical Reasoning 6.2.4
English–Language Arts: Reading 6.2.4

Practice	Assess
❑ SE **Section Review,** p. 251 ❑ **Section Review** ■	❑ SE **Standards Checks,** pp. 245, 246, 248, 249, 250 ❑ TE **Standards Focus,** p. 250 • Assess • Reteach • Re-Assess ❑ **Section Quiz** ■
❑ TE **Universal Access** Debate, p. 245 ❑ TE **Activity** Poster Project, p. 249 ❑ **Critical Thinking**	❑ **Long-Term Projects & Research Ideas** A Whole Lotta Shakin'
❑ TE **Universal Access** Quiz Game, p. 244 ❑ TE **Group Activity** Earthquake Safety, p. 248 ❑ TE **Demonstration** Giant Wave, p. 248 ❑ **Interactive Reader and Study Guide** ❑ **Vocabulary and Section Summary A** ■	❑ TE **Homework** Presentation, p. 246 ❑ TE **Connection to Real Life** Earthquake Kit, p. 249
❑ TE **Universal Access** Earthquakes Around the World, p. 247 ❑ **Interactive Reader and Study Guide** ❑ **Vocabulary and Section Summary A** ■ ❑ **Vocabulary and Section Summary B**	
❑ TE **Universal Access** Earthquake Expectations, p. 245 ❑ TE **Universal Access** Earthquake-Resistant Model, p. 246 ❑ TE **Universal Access** Modeling Effects of an Earthquake, p. 248 ❑ **Interactive Reader and Study Guide**	
❑ **Interactive Reader and Study Guide** ❑ **Directed Reading A** ■ ❑ **Directed Reading B**	❑ TE **Universal Access** Activating Prior Knowledge, p. 249

On Course Mapping Instruction 73

The **California Science Standards** that are covered in each section are identified. Also included are California Math Standards, and California English–Language Arts Standards.

English Learners Students at this level are learning English. These students may be at various learning levels. The resources for these students focus on building language skills and encouraging familiarity with scientific terminology in English.

Struggling Readers Students at this level have difficulty understanding text and recalling or retaining information from text that they have read. Resources for these students extend the reading strategy strand that is built into the Student Edition and Teacher's Edition.

Special Education Students Students at this level may have disabilities, such as visual impairments, hearing impairments, behavioral issues, or delayed development. Resources for these students help teachers adapt teaching methods and activities for the special needs of the students.

Program Resources

Chapter Resource Files (also available on CD-ROM)

The Chapter Resource File booklets contain student worksheets, Teacher Notes, and Answer Keys for the following:

- Directed Reading A
- Directed Reading B
- Vocabulary and Section Summary A
- Vocabulary and Section Summary B
- Reinforcement
- Critical Thinking
- SciLinks® Activity
- Section Reviews
- Chapter Review
- Chapter Pretest
- Section Quizzes
- Chapter Test A
- Chapter Test B
- Chapter Test C
- Performance-Based Assessment
- Standards Assessment
- Datasheets for Explore Activity
- Datasheets for Quick Labs
- Datasheets for Chapter Lab
- Datasheet for Science Skills Activity

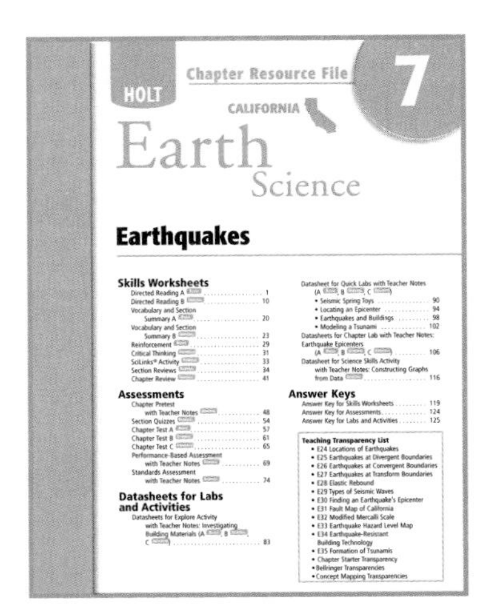

Workbooks

The following workbooks contain student worksheets:

- Study Guide A
- Study Guide B
- Section and Chapter Reviews
- Standards Review Workbook
- Interactive Reader and Study Guide

Spanish Resources

The following resources are also available in Spanish:

- Directed Reading A in the Spanish Study Guide A
- Vocabulary and Section Summary A in the Spanish Study Guide A
- Section Reviews in the Spanish Section and Chapter Reviews workbook
- Chapter Reviews in the Spanish Section and Chapter Reviews workbook
- Section Quizzes in the Spanish Assessments booklet
- Chapter Tests B in the Spanish Assessments booklet
- All lab datasheets on the Lab Generator CD-ROM
- Standards Review Worksheets in the Spanish Standards Review Workbook

Transparencies

- The Teaching Transparencies binder contains all of the Teaching Transparencies for *Holt California Earth Science*, *Holt California Life Science*, and *Holt California Physical Science*.
- Each Additional Transparencies booklet contains the Chapter Starter Transparencies, Bellringer Transparencies, and Concept Mapping Transparencies for one level.
- Standards Review Transparencies are located in the envelope that accompanies the Standards Review Workbook Answer Key.

Lab Generator CD-ROM

The Lab Generator contains all labs from the Student Edition in Spanish and in editable format and contains the following additional labs: Calculator-Based Labs, EcoLabs & Field Activities, Inquiry Labs, Labs You Can Eat, and Long-Term Projects & Research Ideas. Use the Lab Generator to:

- Search for labs by topic, difficulty level, lab duration, or California Science Standard.
- Edit labs to fit classroom needs.
- Develop new labs from our easy-to-use formatted template.
- Save time ordering materials using the Lab Materials QuickList Software.

One-Stop Planner

With the One-Stop Planner, planning and managing lessons has never been easier. This convenient, all-in-one CD-ROM package includes the following time-saving features:

- **All reproducible ancillary worksheets and transparency masters** organized by chapter and section
- **Various Teacher Resources,** including Parent Letters in English and Spanish, the Holt Science Professional Reference for Teachers, and Lab Safety information
- **Holt Calendar Planner**–a tool that allows you to manage your time and resources by the day, week, month, or year
- **Editable lesson plans and lab datasheets**
- **Lab Materials QuickList Software**–a tool to easily create a customizable list of the lab materials you need
- **ExamView® 5.0 Test Generator** (English and Spanish)–a tool that contains thousands of editable questions that are organized by chapter and are correlated to the California Science Standards
- **Holt PuzzlePro®**–an easy way to create crossword puzzles and word searches that make learning vocabulary fun
- **PowerPoint® Resources**–editable presentations that include an outline of chapter content, relevant images from the Student Edition, animated Visual Concepts, and Standards Assessment resources
- **Interactive Teacher's Edition**–the entire Teacher's Edition text, with links to related Teaching Resources

Brain Food Video Quizzes

Brain Food Video Quizzes are engaging, game show–style quizzes that allow students to test their knowledge of chapter content. These videos are available on DVD and VHS.

Lab Videos

Lab Videos make it easier to integrate more experiments into your lessons without the preparation time and costs of a traditional laboratory set-up. These videos demonstrate the end-of-chapter labs and are available on DVD and VHS.

Virtual Investigations CD-ROMs

Virtual Investigations CD-ROMs make it easy for students to practice science skills and perform lab activities in a safe, simulated environment.

1 Pacing

- This chapter should take approximately 6 days to complete.
- Getting Started should take approximately 1 day to complete.

Chapter 1 The Nature of Earth Science

The Big Idea Scientists use careful observations and clear reasoning to understand processes and patterns in nature.

This chapter was designed to cover the California Grade 6 Science Standards about scientific methods (6.7.a, 6.7.d, and 6.7.e) and the California Science Framework requirements regarding safety. This chapter is the first chapter in the *Holt California Earth Science* book because it introduces scientific methods and the thought processes of science. In each of the subsequent chapters, students will benefit from an understanding of the nature of science and the rules of experimentation.

After they have completed this chapter, students will begin a chapter about the tools used by Earth scientists, including maps.

Getting Started

		Teach
	Standards Course of Study	❑ SE **Explore Activity** Planning the Impossible?, p. 7 ❑ **Datasheet B for Explore Activity** Planning the Impossible? ■
Universal Access	Advanced Learners/GATE	❑ **Chapter Starter Transparency** ❑ **Datasheet C for Explore Activity** Planning the Impossible? ■
	Basic Learners	❑ SE **Improving Comprehension,** p. 4 ❑ **Chapter Starter Transparency** ❑ **Datasheet A for Explore Activity** Planning the Impossible? ■
	English Learners	❑ SE **Improving Comprehension,** p. 4 ❑ SE **Unpacking the Standards,** p. 5
	Special Education Students	❑ SE **Improving Comprehension,** p. 4 ❑ SE **Unpacking the Standards,** p. 5
	Struggling Readers	❑ SE **Improving Comprehension,** p. 4

Key

SE Student Edition
TE Teacher's Edition
Chapter Resource File
Workbook
CD or CD-ROM
Transparency
Video
■ Also available in Spanish

All resources listed below are also available on the One-Stop Planner.

The California Science Standards listed below are covered in this chapter:

Investigation and Experimentation

6.7.a Develop a hypothesis.

6.7.d Communicate the steps and results from an investigation in written reports and oral presentations.

6.7.e Recognize whether evidence is consistent with a proposed explanation.

Practice	Assess
❑ **SE Organize Activity** Table Fold, p. 6	❑ **Chapter Pretest**
❑ TE **Using the Unpacked Standards,** p. 5	
❑ TE **Using the Unpacked Standards,** p. 5	
❑ TE **Using the Unpacked Standards,** p. 5	
❑ TE **Using the Unpacked Standards,** p. 5	
❑ TE **Using Other Graphic Organizers,** p. 4	

1 # Pacing • This section should take approximately 1 day to complete.

Section 1 Thinking Like a Scientist

Key Concept Scientific progress is made by asking meaningful questions and conducting careful investigations.

	Teach
Standards Course of Study	❑ **Bellringer Transparency** ❑ **PowerPoint® Resources** ❑ SE **Quick Lab** Using Curiosity to Make Predictions, p. 9 ❑ **Datasheet B for Quick Lab** Using Curiosity to Make Predictions ■
Universal Access: Advanced Learners/GATE	❑ TE **Activity** Talking with Scientists, p. 11 ❑ **Datasheet C for Quick Lab** Using Curiosity to Make Predictions ■
Universal Access: Basic Learners	❑ TE **Activity** Talking with Scientists, p. 11 ❑ **Datasheet A for Quick Lab** Using Curiosity to Make Predictions ■
Universal Access: English Learners	
Universal Access: Special Education Students	❑ TE **Universal Access** Context Clues, p. 12
Universal Access: Struggling Readers	❑ SE **Reading Strategy** Brainstorming, p. 8 ❑ TE **Universal Access** Think Aloud, p. 12

Additional Resources

Holt Lab Generator CD-ROM

Search for any lab by topic, standard, difficulty level, or time. Edit any lab to fit your needs, or create your own labs. Use the Lab Materials QuickList software to customize your lab materials list. Lab datasheets are also available in Spanish on this CD-ROM.

Guided Reading Audio CD Program

The Guided Reading Audio CD Program provides a direct reading of the student text. This resource is helpful to auditory learners and struggling readers. This program is available in English and Spanish.

Key

SE Student Edition	Chapter Resource File	CD or CD-ROM	Video
TE Teacher's Edition	Workbook	Transparency	■ Also available in Spanish

All resources listed below are also available on the One-Stop Planner.

Math: Number Sense 6.1.4

Practice	Assess
❑ SE **Section Review,** p. 15 ❑ **Section Review** ■	❑ SE **Standards Checks,** pp. 11, 13 ❑ TE **Standards Focus,** p. 14 • Assess • Reteach • Re-Assess ❑ **Section Quiz** ■
❑ TE **Universal Access** Scientists as Citizens, p. 13 ❑ TE **Discussion** Consider the Source, p. 12	❑ TE **Research** Scientific Biography, p. 9
❑ TE **Universal Access** Personal Experiences, p. 8 ❑ TE **Discussion** Careers in Science, p. 10 ❑ TE **Activity** Coastal Cleanup, p. 13 ❑ **Interactive Reader and Study Guide** ❑ **Vocabulary and Section Summary A** ■	
❑ TE **Using the Figure** Distorted Science, p. 12 ❑ **Interactive Reader and Study Guide** ❑ **Vocabulary and Section Summary A** ■ ❑ **Vocabulary and Section Summary B**	❑ TE **Group Activity** Ethics Poster, p. 10
❑ TE **Activity** Being Curious, p. 8 ❑ TE **Universal Access** Habits of Mind, p. 10 ❑ **Interactive Reader and Study Guide**	❑ TE **Universal Access** Learning from Experience, p. 11
❑ TE **Universal Access** Understanding Text Features, p. 9 ❑ TE **Reading Strategy** Paired Summarizing, p. 12 ❑ **Interactive Reader and Study Guide** ❑ **Directed Reading A** ■ ❑ **Directed Reading B**	

Reviewing Prior Knowledge

Prepare students to learn how to think like a scientist by having them analyze the volume of two liquids when they are mixed together. See page 8 of the Teacher's Edition.

Subjectivity Students may think that scientists are always objective. To correct this misconception, see page 9 in the Teacher's Edition.

Math Support

Science and math go hand in hand. The Math Practice on page 10 helps students practice math skills in a scientific context.

Section 2 Scientific Methods in Earth Science

Key Concept Scientists conduct careful investigations by following standard methods that allow them to collect data and communicate results.

		Teach
	Standards Course of Study	❑ **Bellringer Transparency** ❑ **PowerPoint® Resources** ❑ **E1** Scientific Methods ❑ **SE Quick Lab** Mapping a Sphere, p. 21 ❑ **Datasheet B for Quick Lab** Mapping a Sphere ■
Universal Access	Advanced Learners/GATE	❑ TE **Universal Access** Describing a Dinosaur's Environment, p. 16 ❑ TE **Discussion** Lack of Proof, p. 20 ❑ **Datasheet C for Quick Lab** Mapping a Sphere ■
Universal Access	Basic Learners	❑ TE **Discussion** Scientific Detectives, p. 16 ❑ TE **Discussion** Kinds of Observations, p. 19 ❑ **Datasheet A for Quick Lab** Mapping a Sphere ■
Universal Access	English Learners	
Universal Access	Special Education Students	❑ TE **Discussion** Recent Discoveries, p. 18 ❑ TE **Universal Access** Section Outline, p. 20
Universal Access	Struggling Readers	❑ **SE Reading Strategy** Outlining, p. 16

Additional Resources

Holt Lab Generator CD-ROM

Search for any lab by topic, standard, difficulty level, or time. Edit any lab to fit your needs, or create your own labs. Use the Lab Materials QuickList software to customize your lab materials list. Lab datasheets are also available in Spanish on this CD-ROM.

Guided Reading Audio CD Program

The Guided Reading Audio CD Program provides a direct reading of the student text. This resource is helpful to auditory learners and struggling readers. This program is available in English and Spanish.

Key

SE Student Edition
TE Teacher's Edition
Chapter Resource File
Workbook
CD or CD-ROM
Transparency
Video
■ Also available in Spanish

All resources listed below are also available on the One-Stop Planner.

Focus on Earth Sciences: 6.7.a, 6.7.d, 6.7.e
English-Language Arts: Reading 6.2.4

Practice	Assess
❑ SE **Section Review,** p. 23 ❑ **Section Review** ■	❑ SE **Standards Checks,** pp. 19, 22 ❑ TE **Standards Focus,** p. 22 • Assess • Reteach • Re-Assess ❑ **Section Quiz** ■
❑ TE **Connection Activity** Communication, p. 21	
❑ **Interactive Reader and Study Guide** ❑ **Vocabulary and Section Summary A** ■ ❑ **Reinforcement Worksheet**	❑ TE **Homework** Using Scientific Methods, p. 16
❑ TE **Group Activity** Solving a Problem, p. 20 ❑ **Interactive Reader and Study Guide** ❑ **Vocabulary and Section Summary A** ■ ❑ **Vocabulary and Section Summary B**	
❑ TE **Universal Access** Rebuilding Dinosaurs, p. 22 ❑ **Interactive Reader and Study Guide**	
❑ TE **Wordwise** Word Connection, p. 18 ❑ TE **Universal Access** SQ4R, p. 18 ❑ **Interactive Reader and Study Guide** ❑ **Directed Reading A** ■ ❑ **Directed Reading B**	

Reviewing Prior Knowledge

Prepare students to learn about scientific methods by having students read the red and blue heads in the section. Then, ask students what they think the section is about. See page 16 of the Teacher's Edition.

MISCONCEPTION ALERT

To Err Can Be Useful Students may confuse hypotheses with predictions. To correct this misconception, see page 20 in the Teacher's Edition.

1 Pacing • This section should take approximately 1 day to complete.

Section 3 Safety in Science

Key Concept Scientific investigations must always be conducted safely.

		Teach
	Standards Course of Study	❑ **Bellringer Transparency** ❑ **PowerPoint® Resources** ❑ **E2** Safety Symbols ❑ SE **Quick Lab** Accident Procedure, p. 28 ❑ **Datasheet B for Quick Lab** Accident Procedure ■
Universal Access	Advanced Learners/GATE	❑ **P3 Link to Physical Science** Using Proper Safety Equipment ❑ **Datasheet C for Quick Lab** Accident Procedure ■
Universal Access	Basic Learners	❑ TE **Discussion** Safety Symbol, p. 25 ❑ **Datasheet A for Quick Lab** Accident Procedure ■
Universal Access	English Learners	❑ TE **Universal Access** Translating Safety Symbols, p. 24 ❑ TE **Discussion** Understanding Procedures, p. 26 ❑ TE **Using the Figure** Safety Equipment, p. 27
Universal Access	Special Education Students	❑ TE **Demonstration** Locating Safety, p. 27
Universal Access	Struggling Readers	❑ SE **Reading Strategy** Graphic Organizer, p. 24 ❑ TE **Universal Access** Application of Safety Symbols, p. 26

Additional Resources

Holt Lab Generator CD-ROM

Search for any lab by topic, standard, difficulty level, or time. Edit any lab to fit your needs, or create your own labs. Use the Lab Materials QuickList software to customize your lab materials list. Lab datasheets are also available in Spanish on this CD-ROM.

Guided Reading Audio CD Program

The Guided Reading Audio CD Program provides a direct reading of the student text. This resource is helpful to auditory learners and struggling readers. This program is available in English and Spanish.

Key

- SE Student Edition
- TE Teacher's Edition
- Chapter Resource File
- Workbook
- CD or CD-ROM
- Transparency
- Video
- ■ Also available in Spanish

All resources listed below are also available on the One-Stop Planner.

Practice	Assess
❑ SE **Section Review,** p. 29 ❑ **Section Review** ■	❑ SE **Standards Check,** p. 26 ❑ TE **Standards Focus,** p. 28 • Assess • Reteach • Re-Assess ❑ **Section Quiz** ■
❑ **Critical Thinking** ❑ **SciLinks Activity**	❑ TE **Universal Access** Design a Habitat, p. 25 ❑ TE **Research** Sports Safety, p. 25
❑ TE **Activity** No Laughing Matter, p. 24 ❑ **Interactive Reader and Study Guide** ❑ **Vocabulary and Section Summary A** ■	❑ TE **Universal Access** Science Safety, p. 25
❑ TE **Group Activity** The Rules, p. 25 ❑ TE **Group Activity** Lab Fashion Show, p. 26 ❑ **Interactive Reader and Study Guide** ❑ **Vocabulary and Section Summary A** ■ ❑ **Vocabulary and Section Summary B**	❑ TE **Group Activity** Safety Fair, p. 27
❑ **Interactive Reader and Study Guide**	
❑ **Interactive Reader and Study Guide** ❑ **Directed Reading A** ■ ❑ **Directed Reading B**	

Reviewing Prior Knowledge

Prepare students to learn about safety by having students study Figure 1. Then, ask students to describe how the biker is being safe. See page 24 of the Teacher's Edition.

Chemical Disposal Students may think that they should pour liquid chemical waste down the sink to dispose of it. To correct this misconception, see page 27 in the Teacher's Edition.

Pacing

- **Chapter Lab, Review, and Assessment should take approximately 2 days to complete.**

Wrapping Up

	Teach
Standards Course of Study	❑ **SE Skills Practice Lab** Using Forensics to Catch a Thief, pp. 30–31 ❑ **Datasheet B for Chapter Lab** Using Forensics to Catch a Thief ■ ❑ **Standards Review Transparency** ■
Universal Access: Advanced Learners/GATE	❑ **Datasheet C for Chapter Lab** Using Forensics to Catch a Thief ■
Universal Access: Basic Learners	❑ **Datasheet A for Chapter Lab** Using Forensics to Catch a Thief ■
Universal Access: English Learners	
Universal Access: Special Education Students	
Universal Access: Struggling Readers	

Additional Resources

SUPER SUMMARY

Have students review the major concepts in this chapter by using the Super Summary that includes the following:

- an outline of important points in the chapter
- flashcards for chapter vocabulary
- an interactive quiz

Go to **go.hrw.com**
Type in the keyword HY7WESS

Performance-Based Assessments

The Chapter Resource File for this chapter contains a hands-on activity that can be used to help assess student progress in a nontraditional format. In the Performance-Based Assessment for this chapter, students analyze the results of an experiment on the Internet.

Key

SE Student Edition
TE Teacher's Edition
Chapter Resource File
Workbook
CD or CD-ROM
Transparency
Video
■ Also available in Spanish

All resources listed below are also available on the One-Stop Planner.

Focus on Earth Sciences: 6.7.a, 6.7.d, 6.7.e
Math: Number Sense 6.1.4
English–Language Arts: Reading 6.2.4; Writing 6.1.1

Practice	Assess
❑ SE **Science Skills Activity** Develop a Hypothesis, p. 32 ❑ **Datasheet for Science Skills Activity** ■ ❑ **Concept Mapping Transparency** ❑ SE **Chapter Review,** pp. 34–35 ❑ **Chapter Review** ■	❑ **Chapter Test B** ■ ❑ SE **Standards Assessment,** pp. 36–37 ❑ **Standards Assessment** ❑ **Standards Review Workbook** ■
❑ SE **Social Studies Activity,** p. 38 ❑ TE **Teaching Strategy** It's Raining Organisms, p. 38	❑ **Chapter Test C** ❑ **Brain Food Video Quiz**
❑ SE **Math Activity,** p. 39	❑ **Chapter Test A** ❑ **Brain Food Video Quiz**
	❑ **Brain Food Video Quiz**
❑ SE **Language Arts Activity,** p. 38 ❑ TE **Universal Access** Tracing Likenesses, p. 32	❑ **Brain Food Video Quiz**
❑ TE **Universal Access** Forming Hypotheses, p. 32	❑ **Brain Food Video Quiz**

Holt Online Assessment

Post tests and quizzes to Holt Online Assessment, an assessment management tool. The system automatically grades the assessments, and you receive students' scores and information about which questions students missed. Holt Online Assessment is available through the Premier Online Edition of *Holt California Earth Science.*

Holt Anthology of Science Fiction

The Holt Anthology of Science Fiction includes thought-provoking stories that are relevant to science instruction. Enhance students' learning by asking them to read a story from the *Holt Anthology of Science Fiction* and to answer questions about what they have read.

Pacing

- This chapter should take approximately 13 days to complete.
- Getting Started should take approximately 1 day to complete.

Chapter 2 Tools of Earth Science

The Big Idea Scientists use a variety of tools, including maps, to perform tests, collect data, and display data.

This chapter was designed to complete the coverage of the California Grade 6 Science Standards about investigation and experimentation (6.7.b, 6.7.c, 6.7.e, and 6.7.f). It follows a chapter about scientific methods and scientific thought processes. This chapter introduces students to the tools and types of models used by scientists. It also describes how to read topographic and geologic maps.

After they have completed this chapter, students will begin a chapter about the transfer of energy and matter through Earth's systems.

Getting Started

		Teach
	Standards Course of Study	❑ SE **Explore Activity** Making and Reading Maps, p. 43 ❑ **Datasheet B for Explore Activity** Making and Reading Maps ■
Universal Access	Advanced Learners/GATE	❑ **Chapter Starter Transparency** ❑ **Datasheet C for Explore Activity** Making and Reading Maps ■
	Basic Learners	❑ SE **Improving Comprehension,** p. 40 ❑ **Chapter Starter Transparency** ❑ **Datasheet A for Explore Activity** Making and Reading Maps ■
	English Learners	❑ SE **Improving Comprehension,** p. 40 ❑ SE **Unpacking the Standards,** p. 41
	Special Education Students	❑ SE **Improving Comprehension,** p. 40 ❑ SE **Unpacking the Standards,** p. 41
	Struggling Readers	❑ SE **Improving Comprehension,** p. 40

Key

SE Student Edition
TE Teacher's Edition
Chapter Resource File
Workbook
CD or CD-ROM
Transparency
Video
■ Also available in Spanish

All resources listed below are also available on the One-Stop Planner.

The California Science Standards listed below are covered in this chapter:

Investigation and Experimentation

6.7.b Select and use appropriate tools and technology (including calculators, computers, balances, spring scales, microscopes, and binoculars) to perform tests, collect data, and display data.

6.7.c Construct appropriate graphs from data and develop qualitative statements about the relationships between variables.

6.7.e Recognize whether evidence is consistent with a proposed explanation.

6.7.f Read a topographic map and a geologic map for evidence provided on the maps and construct and interpret a simple scale map.

Practice	Assess
❑ SE **Organize Activity** Three-Panel Flip Chart, p. 42	❑ **Chapter Pretest**
❑ TE **Words with Multiple Meanings,** p. 41	
❑ TE **Using Other Graphic Organizers,** p. 40	

2 Pacing • This section should take approximately 2 days to complete.

Section 1 Tools and Measurement

Key Concept Scientists must select the appropriate tools to make measurements and collect data, to perform tests, and to analyze data.

		Teach
	Standards Course of Study	❑ **Bellringer Transparency** ❑ **PowerPoint® Resources** ❑ **E3** Measurement Tools ❑ SE **Quick Lab** See for Yourself, p. 45 ❑ **Datasheet B for Quick Lab** See for Yourself ■
Universal Access	Advanced Learners/GATE	❑ **Datasheet C for Quick Lab** See for Yourself ■ ❑ **P56 Link to Physical Science** Finding Volume Using Water Displacement
Universal Access	Basic Learners	❑ TE **Discussion** Computers as Tools, p. 44 ❑ TE **Demonstration** Sweet Surprise, p. 47 ❑ **Datasheet A for Quick Lab** See for Yourself ■
Universal Access	English Learners	❑ TE **Using the Table** Common SI units and Conversion, p. 46
Universal Access	Special Education Students	
Universal Access	Struggling Readers	❑ SE **Reading Strategy** Graphic Organizer, p. 44 ❑ TE **Universal Access** Checking Prior Knowledge, p. 44

Additional Resources

Holt Lab Generator CD-ROM

Search for any lab by topic, standard, difficulty level, or time. Edit any lab to fit your needs, or create your own labs. Use the Lab Materials QuickList software to customize your lab materials list. Lab datasheets are also available in Spanish on this CD-ROM.

Guided Reading Audio CD Program

The Guided Reading Audio CD Program provides a direct reading of the student text. This resource is helpful to auditory learners and struggling readers. This program is available in English and Spanish.

Key

SE Student Edition
TE Teacher's Edition
Chapter Resource File
Workbook
CD or CD-ROM
Transparency
Video
■ Also available in Spanish

All resources listed below are also available on the One-Stop Planner.

Focus on Earth Sciences: 6.7.f
English–Language Arts: Reading 6.2.4

Practice	Assess
❑ SE **Section Review,** p. 49 ❑ **Section Review** ■	❑ SE **Standards Checks,** pp. 47, 48 ❑ TE **Standards Focus,** p. 48 • Assess • Reteach • Re-Assess ❑ **Section Quiz** ■
❑ SE **Reading Strategy** Graphic Organizer, p. 44 ❑ TE **Activity** Note Cards, p. 46 ❑ TE **Universal Access** SI Equivalents, p. 46 ❑ **Interactive Reader and Study Guide** ❑ **Vocabulary and Section Summary A** ■	
❑ TE **Universal Access** Estimates and Measurements, p. 49 ❑ **Interactive Reader and Study Guide** ❑ **Vocabulary and Section Summary A** ■ ❑ **Vocabulary and Section Summary B**	
❑ **Interactive Reader and Study Guide** ❑ TE **Universal Access** Choosing SI Units, p. 47	
❑ TE **Universal Access** Understanding Text Features, p. 46 ❑ **Interactive Reader and Study Guide** ❑ **Directed Reading A** ■ ❑ **Directed Reading B**	

Reviewing Prior Knowledge

Prepare students to learn about tools of measurement by reviewing familiar units of measurement for common measurable quantities. See p. 44 of the Teacher's Edition.

Mass v. Weight Students may have think that mass and weight are the same. To correct this misconception, see p. 45 in the Teacher's Edition.

Math Support

Science and math go hand in hand. The Math Practice item on p.46 and the Math Skills item in the Section Review on p. 49 help students practice math skills in a scientific context.

2 **Pacing** • This section should take approximately 2 days to complete.

Section 2 Models in Science

Key Concept Models are ways of representing real objects or processes to make the natural world easier to understand.

		Teach
	Standards Course of Study	❑ **Bellringer Transparency** ❑ **PowerPoint® Resources** ❑ SE **Quick Lab** See for Yourself, p. 51 ❑ **Datasheet B for Quick Lab** See for Yourself ■
Universal Access	Advanced Learners/GATE	❑ TE **Discussion** Modeling the Past, p. 53 ❑ **Datasheet C for Quick Lab** See for Yourself ■
Universal Access	Basic Learners	❑ TE **Discussion** Modeling the United States, p. 50 ❑ **Datasheet A for Quick Lab** See for Yourself ■
Universal Access	English Learners	❑ TE **Group Activity** Hurricane Models, p. 52
Universal Access	Special Education Students	❑ TE **Universal Access** Modeling Astronomical Theories, p. 54
Universal Access	Struggling Readers	❑ SE **Reading Strategy** Asking Questions, p. 50

Additional Resources

Holt Lab Generator CD-ROM

Search for any lab by topic, standard, difficulty level, or time. Edit any lab to fit your needs, or create your own labs. Use the Lab Materials QuickList software to customize your lab materials list. Lab datasheets are also available in Spanish on this CD-ROM.

Guided Reading Audio CD Program

The Guided Reading Audio CD Program provides a direct reading of the student text. This resource is helpful to auditory learners and struggling readers. This program is available in English and Spanish.

Key

- **SE** Student Edition
- **TE** Teacher's Edition
- Chapter Resource File
- Workbook
- CD or CD-ROM
- Transparency
- Video
- ■ Also available in Spanish

All resources listed below are also available on the One-Stop Planner.

Focus on Earth Sciences: 6.7.b, 6.7.c, 6.7.e
Math: Number Sense 6.1.3

Practice	Assess
❑ SE **Section Review,** p. 55 ❑ **Section Review** ■	❑ SE **Standards Checks,** pp. 50, 52, 54 ❑ TE **Standards Focus,** p. 54 • Assess • Reteach • Re-Assess ❑ **Section Quiz** ■
❑ TE **Universal Access** Scale Models, p. 50	
❑ **Interactive Reader and Study Guide** ❑ **Vocabulary and Section Summary A** ■ ❑ **Reinforcement Worksheet**	
❑ **Interactive Reader and Study Guide** ❑ **Vocabulary and Section Summary A** ■ ❑ **Vocabulary and Section Summary B**	❑ TE **Universal Access** Disproved Theories, p. 53
❑ **Interactive Reader and Study Guide**	
❑ TE **Universal Access** Reading Graphs, p. 51 ❑ TE **Reading Strategy** Paired Summarizing, p. 52 ❑ **Interactive Reader and Study Guide** ❑ **Directed Reading A** ■ ❑ **Directed Reading B**	

Reviewing Prior Knowledge

Prepare students to learn about models in science by reviewing various definitions of the word *model.* See p. 50 of the Teacher's Edition.

Globes and Other Maps Students may have difficulty relating dimensions between different models of Earth. To correct this misconception, see p. 51 in the Teacher's Edition.

Math Support

Science and math go hand in hand. The Math Skills item in the Section Review on p. 55 helps students practice math skills in a scientific context.

2 Pacing • This section should take approximately 2 days to complete.

Section 3 Mapping Earth's Surface

Key Concept Maps are tools that are used to display data about a given area of a physical body.

	Teach
Standards Course of Study	❑ **Bellringer Transparency** ❑ **PowerPoint® Resources** ❑ **E4** Lines of Longitude; Lines of Latitude ❑ **SE Quick Lab** Making a Compass, p. 57 ❑ **Datasheet B for Quick Lab** Making a Compass ■
Universal Access: Advanced Learners/GATE	❑ **Datasheet C for Quick Lab** Making a Compass ■
Universal Access: Basic Learners	❑ **Datasheet A for Quick Lab** Making a Compass ■
Universal Access: English Learners	❑ TE **Using the Figure** Globe Reference Points, p. 56 ❑ TE **Group Activity** Map Skills, p. 58 ❑ TE **Group Activity** Using Directions, p. 58
Universal Access: Special Education Students	
Universal Access: Struggling Readers	❑ SE **Reading Strategy** Summarizing, p. 56 ❑ TE **Reading Strategy** Preview Text Format, p. 59

Additional Resources

Holt Lab Generator CD-ROM

Search for any lab by topic, standard, difficulty level, or time. Edit any lab to fit your needs, or create your own labs. Use the Lab Materials QuickList software to customize your lab materials list. Lab datasheets are also available in Spanish on this CD-ROM.

Guided Reading Audio CD Program

The Guided Reading Audio CD Program provides a direct reading of the student text. This resource is helpful to auditory learners and struggling readers. This program is available in English and Spanish.

Key

SE Student Edition
TE Teacher's Edition
Chapter Resource File
Workbook

CD or CD-ROM
Transparency
Video
■ Also available in Spanish

All resources listed below are also available on the One-Stop Planner.

Focus on Earth Sciences: 6.7.b, 6.7.f
Math: Algebra and Functions 6.2.3
English–Language Arts: Reading 6.2.4

Practice	Assess
❑ SE **Section Review,** p. 63 ❑ **Section Review** ■	❑ SE **Standards Checks,** pp. 56, 59, 61 ❑ TE **Standards Focus,** p. 62 • Assess • Reteach • Re-Assess ❑ **Section Quiz** ■
❑ TE **Universal Access** Chinese Compasses, p. 59 ❑ TE **Connection Activity** Imaginary Places, p. 59 ❑ TE **Universal Access** GIS Simulation, p. 62 ❑ **SciLinks Activity**	
❑ TE **Universal Access** Characteristics of Longitude and Latitude, p. 56 ❑ TE **Activity** Pixels, p. 60 ❑ TE **Universal Access** Researching GIS, p. 62 ❑ **Interactive Reader and Study Guide** ❑ **Vocabulary and Section Summary A** ■	
❑ **Interactive Reader and Study Guide** ❑ **Vocabulary and Section Summary A** ■ ❑ **Vocabulary and Section Summary B**	
❑ TE **Universal Access** Longitude and Latitude, p. 57 ❑ **Interactive Reader and Study Guide**	
❑ **Interactive Reader and Study Guide** ❑ **Directed Reading A** ■ ❑ **Directed Reading B**	❑ TE **Universal Access** Clarifying Understanding, p. 58

Reviewing Prior Knowledge

Prepare students to learn about reading, using, and making maps by discussing how students can find their way to a new place. See p. 56 of the Teacher's Edition.

Map Legends Students may think that all maps are made to the same scale. To correct this misconception, see p. 60 in the Teacher's Edition.

Math Support

Science and math go hand in hand. The Math Practice on p. 61 and the Math Skills item in the Section Review on p. 63 help students practice math skills in a scientific context.

2 Pacing • This section should take approximately 2 days to complete.

Section 4 Maps in Earth Science

Key Concept Topographic and geologic maps include detailed information about Earth's surface and composition.

		Teach
	Standards Course of Study	❑ **Bellringer Transparency** ❑ **PowerPoint® Resources** ❑ **E5** Elevation and Contour Lines ❑ SE **Quick Lab** Modeling Topography, p. 66 ❑ **Datasheet B for Quick Lab** Modeling Topography ■
Universal Access	Advanced Learners/GATE	❑ TE **Connection to Oceanography** Maps of the Ocean Floor, p. 65 ❑ **Datasheet C for Quick Lab** Modeling Topography ■
	Basic Learners	❑ TE **Discussion** Contour Maps, p. 65 ❑ TE **Universal Access** Rules of Contour Lines, p. 66 ❑ TE **Activity** Investigate Your Area, p. 67 ❑ **Datasheet A for Quick Lab** Modeling Topography ■
	English Learners	❑ TE **Using the Figure** Topographic Maps, p. 64 ❑ TE **Using the Figure** El Capitan Topographic Map, p. 67
	Special Education Students	
	Struggling Readers	❑ SE **Reading Strategy** Clarifying Concepts, p. 64

Additional Resources

Holt Lab Generator CD-ROM

Search for any lab by topic, standard, difficulty level, or time. Edit any lab to fit your needs, or create your own labs. Use the Lab Materials QuickList software to customize your lab materials list. Lab datasheets are also available in Spanish on this CD-ROM.

Guided Reading Audio CD Program

The Guided Reading Audio CD Program provides a direct reading of the student text. This resource is helpful to auditory learners and struggling readers. This program is available in English and Spanish.

Key

SE Student Edition
TE Teacher's Edition
Chapter Resource File
Workbook
CD or CD-ROM
Transparency

Video
■ Also available in Spanish

All resources listed below are also available on the One-Stop Planner.

Focus on Earth Sciences: 6.7.f
English–Language Arts: Reading 6.1.1

Practice	Assess
❑ SE **Section Review,** p. 69 ❑ **Section Review** ■	❑ SE **Standards Checks,** pp. 65, 69 ❑ TE **Standards Focus,** p. 62 • Assess • Reteach • Re-Assess ❑ **Section Quiz** ■
❑ **Critical Thinking** ❑ **Inquiry Labs** Looking for Buried Treasure	
❑ **Interactive Reader and Study Guide** ❑ **Vocabulary and Section Summary A** ■	
❑ **Interactive Reader and Study Guide** ❑ **Vocabulary and Section Summary A** ■ ❑ **Vocabulary and Section Summary B**	
❑ **Interactive Reader and Study Guide**	❑ TE **Universal Access** Rules of Contour, p. 67
❑ TE **Universal Access** Understanding Text Features, p. 68 ❑ **Interactive Reader and Study Guide** ❑ **Directed Reading A** ■ ❑ **Directed Reading B**	

Reviewing Prior Knowledge

Prepare students to learn about topographic and geologic maps by reviewing map features. See p. 64 of the Teacher's Edition.

MISCONCEPTION ALERT

Areas of Low Relief Students may think that topographic maps are useful only for studying areas of high relief. To correct this misconception, see p. 65 in the Teacher's Edition.

Math Support

Science and math go hand in hand. The Math Skills item in the Section Review on p. 69 helps students practice math skills in a scientific context.

- **Chapter Lab, Review, and Assessment should take approximately 4 days to complete.**

Wrapping Up

		Teach
	Standards Course of Study	❑ SE **Skills Practice Lab** Topographic Tuber, pp. 70–71 ❑ **Datasheet B for Chapter Lab** Topographic Tuber ■ ❑ **Standards Review Transparency** ■
Universal Access	Advanced Learners/GATE	❑ **Datasheet C for Chapter Lab** Topographic Tuber ■
	Basic Learners	❑ **Datasheet A for Chapter Lab** Topographic Tuber ■
	English Learners	
	Special Education Students	
	Struggling Readers	❑ TE **Universal Access** Reading Maps, p. 72 ❑ TE **Focus on Reading,** p. 73

Additional Resources

SUPER SUMMARY

Have students review the major concepts in this chapter by using the Super Summary that includes the following:

- an outline of important points in the chapter
- flashcards for chapter vocabulary
- an interactive quiz

Go to **go.hrw.com**
Type in the keyword HY7TESS

Performance-Based Assessments

The Chapter Resource File for this chapter contains a hands-on activity that can be used to help assess student progress in a nontraditional format. In the Performance-Based Assessment for this chapter, students invent a standard of measurement.

Key

SE Student Edition
TE Teacher's Edition
Chapter Resource File
Workbook
CD or CD-ROM
Transparency
Video
■ Also available in Spanish

All resources listed below are also available on the One-Stop Planner.

Focus on Earth Sciences: 6.7.b, 6.7.c, 6.7.e, 6.7.f
Math: Algebra and Function 6.2.3
English–Language Arts: Writing 6.1.1

Practice	Assess
❑ **Concept Mapping Transparency** ❑ SE **Chapter Review,** pp. 74–75 ❑ SE **Science Skills Activity** Reading a Topographic Map, p. 72 ❑ **Chapter Review** ■ ❑ **Datasheet for Science Skills Activity** ■	❑ SE **Standards Assessment,** pp. 76–77 ❑ **Standards Assessment** ❑ **Chapter Test B** ■ ❑ **Standards Review Workbook** ■
❑ TE **Identifying Suffixes,** p. 73	❑ **Chapter Test C** ❑ **Brain Food Video Quiz**
❑ TE **Identifying Suffixes,** p. 73 ❑ TE **Activity,** p. 78	❑ SE **Social Studies Activity,** p. 78 ❑ SE **Math Activity,** p. 79 ❑ **Chapter Test A** ❑ **Brain Food Video Quiz** ❑ **Long-Term Projects & Research Ideas** Globe Trotting
❑ SE **Language Arts Activity,** p. 78	❑ **Brain Food Video Quiz**
	❑ **Brain Food Video Quiz**
	❑ **Brain Food Video Quiz**

Holt Online Assessment

Post tests and quizzes to Holt Online Assessment, an assessment management tool. The system automatically grades the assessments, and you receive students' scores and information about which questions students missed. Holt Online Assessment is available through the Premier Online Edition of *Holt California Earth Science.*

Holt Anthology of Science Fiction

The Holt Anthology of Science Fiction includes thought-provoking stories that are relevant to science instruction. Enhance students' learning by asking them to read a story from the *Holt Anthology of Science Fiction* and to answer questions about what they have read.

3 **Pacing**
- This chapter should take approximately 13 days to complete.
- Getting Started should take approximately 1 day to complete.

Chapter 3 Earth's Systems and Cycles

The Big Idea Many phenomena on Earth's surface are affected by the transfer of energy through Earth's systems.

This chapter was designed to cover the California Grade 6 Science Standards about the ways in which energy and matter are transferred through Earth's systems (6.1.b, 6.3.a, 6.3.c, 6.3.d, 6.4.a, 6.4.b, 6.4.c, 6.4.d, 6.5.a, and 6.5.b). This chapter explains that energy from the sun is the major source of energy for phenomena on Earth's surface. This chapter gives students an understanding of the entire Earth and global cycles and gives students the background that they need to study other topics in Earth science.

After they have completed this chapter, students will begin a chapter about Earth's material resources.

Getting Started

		Teach
	Standards Course of Study	❑ SE **Explore Activity** Heat Transfer by Radiation, p. 83 ❑ **Datasheet B for Explore Activity** Heat Transfer by Radiation ■
Universal Access	Advanced Learners/GATE	❑ **Chapter Starter Transparency** ❑ **Datasheet C for Explore Activity** Heat Transfer by Radiation ■
Universal Access	Basic Learners	❑ SE **Improving Comprehension,** p. 80 ❑ **Chapter Starter Transparency** ❑ **Datasheet A for Explore Activity** Heat Transfer by Radiation ■
Universal Access	English Learners	❑ SE **Improving Comprehension,** p. 80 ❑ SE **Unpacking the Standards,** p. 81
Universal Access	Special Education Students	❑ SE **Improving Comprehension,** p. 80 ❑ SE **Unpacking the Standards,** p. 81
Universal Access	Struggling Readers	❑ SE **Improving Comprehension,** p. 80

Key

SE **Student Edition** | **Chapter Resource File** | **CD or CD-ROM** |  **Video**

TE **Teacher's Edition** | **Workbook** | **Transparency** | ■ **Also available in Spanish**

All resources listed below are also available on the One-Stop Planner.

The California Science Standards listed below are covered in this chapter:

Focus on Earth Sciences

6.1.b Students know Earth is composed of several layers: a cold, brittle lithosphere; a hot, convecting mantle; and a dense, metallic core.

6.3.a Students know energy can be carried from one place to another by heat flow or by waves, including water, light and sound waves, or by moving objects.

6.3.c Students know heat flows in solids by conduction (which involves no flow of matter) and in fluids by conduction and by convection (which involves flow of matter).

6.3.d Students know heat energy is also transferred between objects by radiation (radiation can travel through space).

6.4.a Students know the sun is the major source of energy for phenomena on Earth's surface; it powers winds, ocean currents, and the water cycle.

6.4.b Students know solar energy reaches Earth through radiation, mostly in the form of visible light.

6.4.c Students know heat from Earth's interior reaches the surface primarily through convection.

6.4.d Students know convection currents distribute heat in the atmosphere and oceans.

6.5.a Students know energy entering ecosystems as sunlight is transferred by producers into chemical energy through photosynthesis and then from organism to organism through food webs.

6.5.b Students know matter is transferred over time from one organism to others in the food web and between organisms and the physical environment.

Investigation and Experimentation

6.7.a Develop a hypothesis.

6.7.d Communicate the steps and results from an investigation in written reports and oral presentations.

6.7.e Recognize whether evidence is consistent with a proposed explanation.

Practice	Assess
❑ SE **Organize Activity** Four-Corner Fold, p. 82	❑ **Chapter Pretest**
❑ TE **Word Parts,** p. 81	
❑ TE **Using Other Graphic Organizers,** p. 80	

3 **Pacing** • This section should take approximately 2 days to complete.

Section 1 The Earth System

Key Concept Earth is a complex system made up of many smaller systems through which matter and energy are continuously cycled.

	Teach
Standards Course of Study	❑ **Bellringer Transparency** ❑ **PowerPoint® Resources** ❑ **E6** The Layers of Earth ❑ SE **Quick Lab** Rising Heat, p. 86 ❑ **Datasheet B for Quick Lab** Rising Heat ■
Universal Access: Advanced Learners/GATE	❑ **Datasheet C for Quick Lab** Rising Heat ■
Universal Access: Basic Learners	❑ TE **Universal Access** Density Demonstration, p. 87 ❑ **Datasheet A for Quick Lab** Rising Heat ■
Universal Access: English Learners	❑ TE **Using the Figure** Earth's Layers, p. 85 ❑ TE **Wordwise** Spheres on Earth, p. 85
Universal Access: Special Education Students	❑ TE **Demonstration** Heating the Surface, p. 87 ❑ TE **Universal Access** Comparing Wedges, p. 85
Universal Access: Struggling Readers	❑ SE **Reading Strategy** Graphic Organizer, p. 84

Key

SE Student Edition
TE Teacher's Edition
Chapter Resource File
Workbook
CD or CD-ROM
Transparency
Video
■ Also available in Spanish

All resources listed below are also available on the One-Stop Planner.

Focus on Earth Sciences: 6.1.b, 6.3.c, 6.4.a, 6.4.b, 6.4.d, 6.5.a, 6.5.b
Math: Number Sense 6.1.4

Practice	Assess
❑ SE **Section Review,** p. 89 ❑ **Section Review** ■	❑ SE **Standards Checks,** pp. 85, 86, 87, 88, 89 ❑ TE **Standards Focus,** p. 88 • Assess • Reteach • Re-Assess ❑ **Section Quiz** ■
❑ TE **Universal Access** Researching the Atmosphere, p. 86	
❑ **Interactive Reader and Study Guide** ❑ **Vocabulary and Section Summary A** ■	
❑ TE **Group Activity** Tectonic Plates, p. 84 ❑ TE **Universal Access** Vocabulary Focus, p. 84 ❑ **Interactive Reader and Study Guide** ❑ **Vocabulary and Section Summary A** ■ ❑ **Vocabulary and Section Summary B**	
❑ **Interactive Reader and Study Guide**	
❑ TE **Universal Access** Monitoring Understanding of Key Terms, p. 85 ❑ TE **Wordwise** Reviewing Convection, p. 87 ❑ **Interactive Reader and Study Guide** ❑ **Directed Reading A** ■ ❑ **Directed Reading B**	

 • This section should take approximately 2 days to complete.

Section 2 Heat and Energy

Key Concept Heat flows in a predictable way from warmer objects to cooler objects until all of the objects are at the same temperature.

	Teach
Standards Course of Study	❑ **Bellringer Transparency** ❑ **PowerPoint® Resources** ❑ SE **Quick Lab** Heat Exchange, p. 95 ❑ **Datasheet B for Quick Lab** Heat Exchange ■
Universal Access: Advanced Learners/GATE	❑ TE **Discussion** Everyday Thermal Expansion, p. 91 ❑ **Datasheet C for Quick Lab** Heat Exchange ■
Universal Access: Basic Learners	❑ TE **Demonstration** Thermal Expansion, p. 90 ❑ TE **Universal Access** Heat Transfer, p. 93 ❑ TE **Discussion** Transferred Energy, p. 94 ❑ **Datasheet A for Quick Lab** Heat Exchange ■
Universal Access: English Learners	
Universal Access: Special Education Students	❑ TE **Universal Access** Modeling Kinetic Energy, p. 90 ❑ TE **Universal Access** Cool-Hand Students, p. 92 ❑ TE **Demonstration** Conduction Materials, p. 94
Universal Access: Struggling Readers	❑ SE **Reading Strategy** Prediction Guide, p. 90 ❑ TE **Universal Access** Structural Patterns, p. 90 ❑ TE **Reading Strategy** Prediction Guide, p. 93

Additional Resources

Holt Lab Generator CD-ROM

Search for any lab by topic, standard, difficulty level, or time. Edit any lab to fit your needs, or create your own labs. Use the Lab Materials QuickList software to customize your lab materials list. Lab datasheets are also available in Spanish on this CD-ROM.

Guided Reading Audio CD Program

The Guided Reading Audio CD Program provides a direct reading of the student text. This resource is helpful to auditory learners and struggling readers. This program is available in English and Spanish.

Key

SE Student Edition	Chapter Resource File	CD or CD-ROM	Video
TE Teacher's Edition	Workbook	Transparency	■ Also available in Spanish

All resources listed below are also available on the One-Stop Planner.

Focus on Earth Sciences: 6.3.a, 6.3.c, 6.3.d

Practice	Assess
❑ SE **Section Review,** p. 97 ❑ **Section Review** ■	❑ SE **Standards Checks,** pp. 93, 95 ❑ TE **Standards Focus,** p. 88 • Assess • Reteach • Re-Assess ❑ **Section Quiz** ■
❑ TE **Universal Access** Thermal Conductivity, p. 95	❑ **Calculator-Based Labs** A Hot and Cool Lab
❑ **Interactive Reader and Study Guide** ❑ **Vocabulary and Section Summary A** ■ ❑ **Reinforcement Worksheet**	
❑ TE **Universal Access** Concentrating on Vocabulary, p. 91 ❑ **Interactive Reader and Study Guide** ❑ **Vocabulary and Section Summary A** ■ ❑ **Vocabulary and Section Summary B**	
❑ TE **Activity** Energy Comics, p. 92 ❑ **Interactive Reader and Study Guide**	
❑ TE **Universal Access** Outlining, p. 94 ❑ **Interactive Reader and Study Guide** ❑ **Directed Reading A** ■ ❑ **Directed Reading B**	

Reviewing Prior Knowledge

Prepare students to learn about kinetic energy and temperature by having students study Figure 1. See page 90 of the Teacher's Edition.

The Meaning of Heat Students may think that heat means warmth. To correct this misconception, see page 92 in the Teacher's Edition.

MISCONCEPTION ALERT

The Meaning of Heat Students may think that "cold" is transferred from a cool object to a warmer object. To correct this misconception, see page 93 in the Teacher's Edition.

3 Pacing • This section should take approximately 2 days to complete.

Section 3 The Cycling of Energy

Key Concept Various heat-exchange systems work in the Earth system and affect phenomena on Earth's surface.

		Teach
	Standards Course of Study	❑ **Bellringer Transparency** ❑ **PowerPoint® Resources** ❑ **E7** Convection in the Geosphere ❑ **SE Quick Lab** Modeling Convection, p. 102 ❑ **Datasheet B for Quick Lab** Modeling Convection ■
Universal Access	Advanced Learners/GATE	❑ **Datasheet C for Quick Lab** Modeling Convection ■
Universal Access	Basic Learners	❑ TE **Using the Figure** Electromagnetic Spectrum, p. 99 ❑ TE **Using the Figure** Ocean Currents, p. 100 ❑ **Datasheet A for Quick Lab** Modeling Convection ■
Universal Access	English Learners	❑ TE **Universal Access** Demonstrating Heat Flow, p. 98
Universal Access	Special Education Students	
Universal Access	Struggling Readers	❑ **SE Reading Strategy** Graphic Organizer, p. 98 ❑ TE **Universal Access** Recognizing References, p. 103

Additional Resources

Holt Lab Generator CD-ROM

Search for any lab by topic, standard, difficulty level, or time. Edit any lab to fit your needs, or create your own labs. Use the Lab Materials QuickList software to customize your lab materials list. Lab datasheets are also available in Spanish on this CD-ROM.

Guided Reading Audio CD Program

The Guided Reading Audio CD Program provides a direct reading of the student text. This resource is helpful to auditory learners and struggling readers. This program is available in English and Spanish.

Key

SE Student Edition
TE Teacher's Edition
Chapter Resource File
Workbook
CD or CD-ROM
Transparency
Video
■ Also available in Spanish

All resources listed below are also available on the One-Stop Planner.

Focus on Earth Sciences: 6.3.a, 6.3.c, 6.3.d, 6.4.a, 6.4.b, 6.4.c, 6.4.d
Math: Number Sense 6.1.4

Practice	Assess
❑ SE **Section Review,** p. 103 ❑ **Section Review** ■	❑ SE **Standards Checks,** pp. 98, 99, 101, 102 ❑ TE **Standards Focus,** p. 102 • Assess • Reteach • Re-Assess ❑ **Section Quiz** ■
	❑ TE **Homework** El Niño, p. 100 ❑ TE **Universal Access** Convection Animation, p. 101
❑ TE **Universal Access** Mnemonic Device, p. 99 ❑ **Interactive Reader and Study Guide** ❑ **Vocabulary and Section Summary A** ■	❑ **Labs You Can Eat** Famous Rock Groups
❑ **Interactive Reader and Study Guide** ❑ **Vocabulary and Section Summary A** ■ ❑ **Vocabulary and Section Summary B**	
❑ TE **Universal Access** Reference Entry, p. 100 ❑ TE **Activity** Energy Cycles, p. 98 ❑ **Interactive Reader and Study Guide**	
❑ TE **Reading Strategy** Reading Organizer, p. 99 ❑ **Interactive Reader and Study Guide** ❑ **Directed Reading A** ■ ❑ **Directed Reading B**	

Reviewing Prior Knowledge

Prepare students to learn about heat flow by asking students what they already know about how heat flows. See page 98 of the Teacher's Edition.

Convection Currents Students may think that convection currents in the geosphere move just as fast as convection currents in the atmosphere or ocean do. To correct this misconception, see page 101 in the Teacher's Edition.

Math Support

Science and math go hand in hand. The Math Practice on p. 100 helps students practice math skills in a scientific context.

3 **Pacing** • This section should take approximately 2 days to complete.

Section 4 The Cycling of Matter

Key Concept Over time, matter—such as rock, water, carbon, and nitrogen—is transferred between organisms and the physical environment.

		Teach
	Standards Course of Study	❑ **Bellringer Transparency** ❑ **PowerPoint® Resources** ❑ **E8** The Rock Cycle ❑ SE **Quick Lab** Modeling the Water Cycle, p. 108 ❑ **Datasheet B for Quick Lab** Modeling the Water Cycle ■
Universal Access	Advanced Learners/GATE	❑ **P39 Link to Physical Science** Three Kinds of Carbon Backbones ❑ **Datasheet C for Quick Lab** Modeling the Water Cycle ■
	Basic Learners	❑ TE **Using the Figure** Geologic Eras, p. 104 ❑ TE **Using the Figure** Reviewing the Rock Cycle, p. 105 ❑ TE **Demonstration** Metamorphism, p. 107 ❑ TE **Discussion** The Carbon Cycle, p. 109 ❑ **Datasheet A for Quick Lab** Modeling the Water Cycle ■
	English Learners	
	Special Education Students	❑ TE **Demonstration** Illustrating the Rock Cycle, p. 104 ❑ TE **Universal Access** I Am Igneous, p. 105 ❑ TE **Activity** Rock Cycle Diagram, p. 105 ❑ TE **Universal Access** Sorting Cycles, p. 108
	Struggling Readers	❑ SE **Reading Strategy** Outlining, p. 104 ❑ TE **Universal Access** Understanding Diagrams, p. 106

Key

SE Student Edition
TE Teacher's Edition
Chapter Resource File
Workbook
CD or CD-ROM
Transparency
Video
■ Also available in Spanish

All resources listed below are also available on the One-Stop Planner.

Focus on Earth Sciences: 6.4.a, 6.5.a, 6.5.b
English–Language Arts: Reading 6.2.4

Practice	Assess
❑ SE **Section Review,** p. 111 ❑ **Section Review** ■	❑ SE **Standards Checks,** pp. 108, 110 ❑ TE **Standards Focus,** p. 110 • Assess • Reteach • Re-Assess ❑ **Section Quiz** ■
❑ **Critical Thinking** ❑ **SciLinks Activity**	❑ TE **Universal Access** Guide to Rocks, p. 107 ❑ TE **Homework** Carbon Dioxide and Global Warming, p. 109 ❑ **Calculator-Based Labs** The Power of the Sun
❑ TE **Universal Access** Water Cycle Poster, p. 108 ❑ **Interactive Reader and Study Guide** ❑ **Vocabulary and Section Summary A** ■	
❑ TE **Group Activity** Describing Rocks, p. 106 ❑ TE **Group Activity** Classifying Rocks, p. 107 ❑ TE **Universal Access** Writing about Cycles, p. 110 ❑ **Interactive Reader and Study Guide** ❑ **Vocabulary and Section Summary A** ■ ❑ **Vocabulary and Section Summary B**	
❑ **Interactive Reader and Study Guide**	
❑ **Interactive Reader and Study Guide** ❑ **Directed Reading A** ■ ❑ **Directed Reading B**	

3 Pacing

- Chapter Lab, Review, and Assessment should take approximately 2 days to complete.

Wrapping Up

	Teach
Standards Course of Study	❑ SE **Inquiry Lab** Stop the Energy Transfer, pp. 112–113 ❑ **Datasheet B for Chapter Lab** Stop the Energy Transfer ■ ❑ **Standards Review Transparency** ■
Universal Access: Advanced Learners/GATE	❑ **Datasheet C for Chapter Lab** Stop the Energy Transfer ■
Universal Access: Basic Learners	❑ **Datasheet A for Chapter Lab** Stop the Energy Transfer ■
Universal Access: English Learners	
Universal Access: Special Education Students	
Universal Access: Struggling Readers	❑ TE **Universal Access** Practicing Structural Patterns, p. 114

Additional Resources

SUPER SUMMARY

Have students review the major concepts in this chapter by using the Super Summary that includes the following:

- an outline of important points in the chapter
- flashcards for chapter vocabulary
- an interactive quiz

Go to **go.hrw.com**
Type in the keyword HY7ESCS

Performance-Based Assessments

The Chapter Resource File for this chapter contains a hands-on activity that can be used to help assess student progress in a nontraditional format. In the Performance-Based Assessment for this chapter, students model radiation and convection.

Key

SE Student Edition
TE Teacher's Edition
Chapter Resource File
Workbook
CD or CD-ROM
Transparency
Video
■ Also available in Spanish

All resources listed below are also available on the One-Stop Planner.

Focus on Earth Sciences: 6.1.b, 6.3.a, 6.3.c, 6.3.d, 6.4.a, 6.4.b, 6.4.c, 6.4.d, 6.5.a, 6.5.b
Math: Number Sense 6.1.4; Algebra and Functions 6.2.1
English–Language Arts: Reading 6.2.4; Writing 6.1.3, 6.2.1

Practice	Assess
❑ SE **Science Skills Activity** Communicating Results, p. 114 ❑ **Datasheet for Science Skills Activity** ■ ❑ **Concept Mapping Transparency** ❑ SE **Chapter Review,** pp. 116–117 ❑ **Chapter Review** ■	❑ **Chapter Test B** ■ ❑ SE **Standards Assessment,** pp. 118–119 ❑ **Standards Assessment** ❑ **Standards Review Workbook** ■
	❑ SE **Social Studies Activity,** p. 121 ❑ **Chapter Test C** ❑ **Brain Food Video Quiz**
❑ SE **Math Activity,** p. 120	❑ **Chapter Test A** ❑ **Brain Food Video Quiz**
❑ TE **Identifying Prefixes** Atmospheric Activity, p. 115 ❑ SE **Language Arts Activity,** p. 120	❑ TE **Focus on Writing** Relating Concepts, p. 115 ❑ **Brain Food Video Quiz**
❑ TE **Universal Access** Presentation Accommodations, p. 114	❑ **Brain Food Video Quiz**
❑ TE **Identifying Prefixes** Atmospheric Activity, p. 115	❑ **Brain Food Video Quiz**

Holt Online Assessment

Post tests and quizzes to Holt Online Assessment, an assessment management tool. The system automatically grades the assessments, and you receive students' scores and information about which questions students missed. Holt Online Assessment is available through the Premier Online Edition of *Holt California Earth Science.*

Holt Anthology of Science Fiction

The Holt Anthology of Science Fiction includes thought-provoking stories that are relevant to science instruction. Enhance students' learning by asking them to read a story from the *Holt Anthology of Science Fiction* and to answer questions about what they have read.

4

Pacing

- This chapter should take approximately 10 days to complete.
- Getting Started should take approximately 1 day to complete.

Chapter 4 Material Resources

The Big Idea Material resources differ in amounts, distribution, usefulness, and the time required for their formation.

This chapter was designed to cover the California Grade 6 Science Standards about natural resources (6.6.b and 6.6.c). The chapter also covers the discussion of environmental science and the wise use of resources referred to in Category 1, Criterion 11, in the Criteria for Evaluating Instructional Materials in Science. It follows a chapter that introduced Earth's systems and cycles of energy and matter on Earth. This chapter explains the difference between renewable and nonrenewable resources. This chapter also describes different material resources and how they are used to make common objects.

After they have completed this chapter, students will begin a chapter about energy resources.

Getting Started

		Teach
	Standards Course of Study	❑ SE **Explore Activity** What Is Your Classroom Made Of?, p. 127 ❑ **Datasheet B for Explore Activity** What Is Your Classroom Made Of? ■
Universal Access	Advanced Learners/GATE	❑ **Chapter Starter Transparency** ❑ **Datasheet C for Explore Activity** What Is Your Classroom Made Of? ■
Universal Access	Basic Learners	❑ SE **Improving Comprehension,** p. 124 ❑ **Chapter Starter Transparency** ❑ **Datasheet A for Explore Activity** What Is Your Classroom Made Of? ■
Universal Access	English Learners	❑ SE **Improving Comprehension,** p. 124 ❑ SE **Unpacking the Standards,** p. 125
Universal Access	Special Education Students	❑ SE **Improving Comprehension,** p. 124 ❑ SE **Unpacking the Standards,** p. 125
Universal Access	Struggling Readers	❑ SE **Improving Comprehension,** p. 124

Key

SE Student Edition
TE Teacher's Edition
Chapter Resource File
Workbook
CD or CD-ROM
Transparency
Video
■ Also available in Spanish

All resources listed below are also available on the One-Stop Planner.

The California Science Standards listed below are covered in this chapter:

Focus on Earth Sciences

6.6.b Students know different natural energy and material resources, including air, soil, rocks, minerals, petroleum, fresh water, wildlife, and forests, and know how to classify them as renewable or nonrenewable.

6.6.c Students know the natural origin of the materials used to make common objects.

Investigation and Experimentation

6.7.b Select and use appropriate tools and technology (including calculators, computers, balances, spring scales, microscopes, and binoculars) to perform tests, collect data, and display data.

6.7.f Read a topographic map and a geologic map for evidence provided on the maps and construct and interpret a simple scale map.

Practice	Assess
❑ SE **Organize Activity** Two-Panel Flip Chart, p. 126	❑ **Chapter Pretest**
❑ TE **Academic Vocabulary,** p. 125	
❑ TE **Using Other Graphic Organizers,** p. 124	

4 # Pacing • This section should take approximately 1 day to complete.

Section 1 Natural Resources

Key Concept Different energy and material resources can be classified as renewable or nonrenewable.

	Teach
Standards Course of Study	❑ **Bellringer Transparency** ❑ **PowerPoint® Resources** ❑ SE **Quick Lab** Renewable or Not?, p. 129 ❑ **Datasheet B for Quick Lab** Renewable or Not? ■
Universal Access: Advanced Learners/GATE	❑ TE **Wordwise** Prefix and Root, p. 131 ❑ **Datasheet C for Quick Lab** Renewable or Not? ■
Universal Access: Basic Learners	❑ TE **Universal Access** Picturing Resources, p. 128 ❑ **Datasheet A for Quick Lab** Renewable or Not? ■
Universal Access: English Learners	❑ TE **Wordwise** Prefix and Root, p. 131 ❑ TE **Using the Figure** Recycling, p. 131
Universal Access: Special Education Students	❑ TE **Universal Access** Energy Conservation Survey, p. 130
Universal Access: Struggling Readers	❑ SE **Reading Strategy** Clarifying Concepts, p. 128 ❑ TE **Universal Access** Identifying Structural Patterns, p. 128 ❑ TE **Using the Figure** Recycling, p. 131

Additional Resources

Holt Lab Generator CD-ROM

Search for any lab by topic, standard, difficulty level, or time. Edit any lab to fit your needs, or create your own labs. Use the Lab Materials QuickList software to customize your lab materials list. Lab datasheets are also available in Spanish on this CD-ROM.

Guided Reading Audio CD Program

The Guided Reading Audio CD Program provides a direct reading of the student text. This resource is helpful to auditory learners and struggling readers. This program is available in English and Spanish.

Key

SE Student Edition
TE Teacher's Edition
Chapter Resource File
Workbook
CD or CD-ROM
Transparency
Video
■ Also available in Spanish

Focus on Earth Sciences: 6.6.b
English–Language Arts: Reading 6.1.1

Practice	Assess
❑ SE **Section Review,** p. 131 ❑ **Section Review** ■	❑ SE **Standards Checks,** pp. 129, 131 ❑ TE **Standards Focus,** p. 130 • Assess • Reteach • Re-Assess ❑ **Section Quiz** ■
	❑ TE **Universal Access** Making a Conservation Flow Chart, p. 130
❑ **Interactive Reader and Study Guide** ❑ **Vocabulary and Section Summary A** ■ ❑ **Reinforcement Worksheet**	
❑ **Interactive Reader and Study Guide** ❑ **Vocabulary and Section Summary A** ■ ❑ **Vocabulary and Section Summary B**	
❑ **Interactive Reader and Study Guide**	
❑ TE **Activity** Role Playing, p. 128 ❑ **Interactive Reader and Study Guide** ❑ **Directed Reading A** ■ ❑ **Directed Reading B**	

Reviewing Prior Knowledge

Prepare students to learn about natural resources by reviewing the things human beings need to survive. See page 128 of the Teacher's Edition.

Endangered Renewable Resources
Students may think that any resource that is renewable will be available forever. To correct this misconception, see page 129 in the Teacher's Edition.

Math Support

Science and math go hand in hand.
The Math Skills item in the Section Review on page 131 helps students practice math skills in a scientific context.

Section 2 Rock and Mineral Resources

Key Concept Minerals and ores are important sources of materials that are used to make common objects.

	Teach
Standards Course of Study	❑ **Bellringer Transparency** ❑ **PowerPoint® Resources** ❑ **P25 Link to Physical Science** Properties of Alkali Metals ❑ SE **Quick Lab** Chocolate Ore, p. 135 ❑ **Datasheet B for Quick Lab** Chocolate Ore ■
Universal Access: Advanced Learners/GATE	❑ TE **Activity** Mineral Search, p. 132 ❑ **Datasheet C for Quick Lab** Chocolate Ore ■
Universal Access: Basic Learners	❑ TE **Universal Access** Diagramming Mine Types, p. 135 ❑ **Datasheet A for Quick Lab** Chocolate Ore ■
Universal Access: English Learners	❑ TE **Activity** Mineral Search, p. 132 ❑ TE **Universal Access** Write About It!, p. 135
Universal Access: Special Education Students	❑ TE **Universal Access** Paired Discussions, p. 132
Universal Access: Struggling Readers	❑ SE **Reading Strategy** Graphic Organizer, p. 132 ❑ TE **Activity** Mineral Search, p. 132 ❑ TE **Universal Access** Using Context Clues, p. 134

Additional Resources

Holt Lab Generator CD-ROM

Search for any lab by topic, standard, difficulty level, or time. Edit any lab to fit your needs, or create your own labs. Use the Lab Materials QuickList software to customize your lab materials list. Lab datasheets are also available in Spanish on this CD-ROM.

Guided Reading Audio CD Program

The Guided Reading Audio CD Program provides a direct reading of the student text. This resource is helpful to auditory learners and struggling readers. This program is available in English and Spanish.

Key

SE Student Edition
TE Teacher's Edition

Chapter Resource File
Workbook

CD or CD-ROM
Transparency

Video
■ Also available in Spanish

All resources listed below are also available on the One-Stop Planner.

Focus on Earth Sciences: 6.6.b, 6.6.c
Math: Number Sense 6.1.4, 6.2.1
English–Language Arts: Reading 6.2.4

Practice	Assess
❑ SE **Section Review,** p. 137 ❑ **Section Review** ■	❑ SE **Standards Checks,** pp. 132, 134, 137 ❑ TE **Standards Focus,** p. 136 • Assess • Reteach • Re-Assess ❑ **Section Quiz** ■
❑ **Long-Term Projects & Research Ideas** Home-Grown Crystals	❑ TE **Universal Access** Minerals Chart, p. 133 ❑ TE **Connection to Social Studies** The History of Mining Communities, p. 134 ❑ TE **Connection to Real World** The Costs of Aluminum, p. 134 ❑ **Long-Term Projects & Research Ideas** What's Yours Is Mined
❑ **Interactive Reader and Study Guide** ❑ **Vocabulary and Section Summary A** ■	
❑ **Interactive Reader and Study Guide** ❑ **Vocabulary and Section Summary A** ■ ❑ **Vocabulary and Section Summary B**	❑ TE **Homework** Report, p. 133
❑ **Interactive Reader and Study Guide**	
❑ **Interactive Reader and Study Guide** ❑ **Directed Reading A** ■ ❑ **Directed Reading B**	

Reviewing Prior Knowledge

Prepare students to learn about rock and mineral resources by using the red and blue section heads to spark a discussion about what students already know about minerals and mining. See page 132 of the Teacher's Edition.

Is It a Gem? Students may think that *gem* and *mineral* are interchangeable terms. To correct this misconception, see page 133 in the Teacher's Edition.

Math Support

Science and math go hand in hand. The Math Practice on page 134 and the Math Skills item in the Section Review on page 137 help students practice math skills in a scientific context.

Section 3 Using Material Resources

Key Concept A variety of natural resources are used to make common objects.

	Teach
Standards Course of Study	❑ **Bellringer Transparency** ❑ **PowerPoint® Resources** ❑ **E9** Meeting Human Needs with Natural Resources ❑ SE **Quick Lab** Products from Plants, p. 141 ❑ **Datasheet B for Quick Lab** Products from Plants ■ ❑ **E10** The Production of Paper
Universal Access: Advanced Learners/GATE	❑ TE **Discussion** Mining on Other Planets, p. 139 ❑ TE **Wordwise** Root Words, p. 139 ❑ TE **Connection to Life Science** Resource Use and Global Warming, p. 139 ❑ **Datasheet C for Quick Lab** Products from Plants ■
Universal Access: Basic Learners	❑ TE **Universal Access** Environmental Art, p. 139 ❑ TE **Connection to Life Science** Resource Use and Global Warming, p. 139 ❑ **Datasheet A for Quick Lab** Products from Plants ■
Universal Access: English Learners	❑ TE **Group Activity** Resource Use and Cost Analysis, p. 138 ❑ TE **Discussion** Mining on Other Planets, p. 139 ❑ TE **Wordwise** Root Words, p. 139 ❑ TE **Connection to Life Science** Resource Use and Global Warming, p. 139 ❑ TE **Using the Figure** Building Materials, p. 140 ❑ TE **Universal Access** Natural Resources and Daily Life, p. 140
Universal Access: Special Education Students	❑ TE **Universal Access** Natural Resources Concept Maps, p. 141
Universal Access: Struggling Readers	❑ SE **Reading Strategy** Asking Questions, p. 138 ❑ TE **Universal Access** Identifying Structural Patterns, p. 138 ❑ TE **Connection to Life Science** Resource Use and Global Warming, p. 139 ❑ TE **Using the Figure** Building Materials, p. 140 ❑ TE **Universal Access** Drawing Conclusions, p. 142

Key

SE Student Edition
TE Teacher's Edition
Chapter Resource File
Workbook
CD or CD-ROM
Transparency
Video
■ Also available in Spanish

All resources listed below are also available on the One-Stop Planner.

Focus on Earth Sciences: 6.6.b, 6.6.c
Math: Number Sense 6.1.4

Practice	Assess
❑ SE **Section Review,** p. 143 ❑ **Section Review** ■	❑ SE **Standards Checks,** pp. 139, 141, 142 ❑ TE **Standards Focus,** p. 142 • Assess • Reteach • Re-Assess ❑ **Section Quiz** ■
❑ **Critical Thinking** ❑ **SciLinks Activity**	❑ TE **Cultural Awareness** Poster Project, p. 140 ❑ TE **Universal Access** Researching Economic Costs, p. 140 ❑ **Long-Term Projects & Research Ideas** Let's Talk Trash ❑ **Long-Term Projects & Research Ideas** Your Very Own Underwater Theme Park
❑ **Interactive Reader and Study Guide** ❑ **Vocabulary and Section Summary A** ■	
❑ **Interactive Reader and Study Guide** ❑ **Vocabulary and Section Summary A** ■ ❑ **Vocabulary and Section Summary B**	
❑ **Interactive Reader and Study Guide**	
❑ **Interactive Reader and Study Guide** ❑ **Directed Reading A** ■ ❑ **Directed Reading B**	

Pacing

- Chapter Lab, Review, and Assessment should take approximately 4 days to complete.

Wrapping Up

	Teach
Standards Course of Study	❑ SE **Skills Practice Lab** Natural Resources Used at Lunch, pp. 144–145 ❑ **Datasheet B for Chapter Lab** Natural Resources Used at Lunch ■ ❑ **Standards Review Transparency** ■
Universal Access: Advanced Learners/GATE	❑ **Datasheet C for Chapter Lab** Natural Resources Used at Lunch ■
Universal Access: Basic Learners	❑ **Datasheet A for Chapter Lab** Natural Resources Used at Lunch ■
Universal Access: English Learners	❑ TE **Identifying Prefixes and Suffixes** Vocabulary Affixes, p. 147
Universal Access: Special Education Students	❑ TE **Universal Access** Patterns and Textures, p. 146 ❑ TE **Universal Access** Distinguishing Text Features, p. 146
Universal Access: Struggling Readers	

Additional Resources

SUPER SUMMARY

Have students review the major concepts in this chapter by using the Super Summary that includes the following:

- an outline of important points in the chapter
- flashcards for chapter vocabulary
- an interactive quiz

Go to **go.hrw.com**
Type in the keyword HY7MARS

Performance-Based Assessments

The Chapter Resource File for this chapter contains a hands-on activity that can be used to help assess student progress in a nontraditional format. In the Performance-Based Assessment for this chapter, students learn about mining processes.

Key

- **SE** Student Edition
- **TE** Teacher's Edition
- Chapter Resource File
- Workbook
- CD or CD-ROM
- Transparency
- Video
- ■ Also available in Spanish

All resources listed below are also available on the One-Stop Planner.

Focus on Earth Sciences: 6.6.b, 6.6.c, 6.7.b, 6.7.f
Math: Statistics, Data Analysis, and Probability 6.1.1; Number Sense 6.1.4, 6.2.1
English–Language Arts: Writing 6.1.3, 6.2.3

Practice	Assess
❑ **SE Science Skills Activity** Reading a Geologic Map, p. 146 ❑ **Datasheet for Science Skills Activity** ■ ❑ **Concept Mapping Transparency** ❑ **SE Chapter Review,** pp. 148–149 ❑ **Chapter Review** ■	❑ **Chapter Test B** ■ ❑ **SE Standards Assessment,** pp. 150–151 ❑ **Standards Assessment** ❑ **Standards Review Workbook** ■
❑ **SE Math Activity,** p. 152 ❑ **TE Connection to Math** Carat Conversions, p. 153 ❑ **TE Activity** Local Minerals, p. 153	❑ **TE Focus on Writing** Public Service Announcement, p. 147 ❑ **SE Language Arts Activity,** p. 152 ❑ **SE Social Studies Activity,** p. 152 ❑ **Chapter Test C** ❑ **Brain Food Video Quiz**
	❑ **Chapter Test A** ❑ **Brain Food Video Quiz**
❑ **TE Activity** Local Minerals, p. 153	❑ **TE Focus on Writing** Public Service Announcement, p. 147 ❑ **Brain Food Video Quiz**
	❑ **Brain Food Video Quiz**
	❑ **Brain Food Video Quiz**

Holt Online Assessment

Post tests and quizzes to Holt Online Assessment, an assessment management tool. The system automatically grades the assessments, and you receive students' scores and information about which questions students missed. Holt Online Assessment is available through the Premier Online Edition of *Holt California Earth Science.*

Holt Anthology of Science Fiction

The Holt Anthology of Science Fiction includes thought-provoking stories that are relevant to science instruction. Enhance students' learning by asking them to read a story from the *Holt Anthology of Science Fiction* and to answer questions about what they have read.

5 **Pacing**
- This chapter should take approximately 8 days to complete.
- Getting Started should take approximately 1 day to complete.

Chapter 5 Energy Resources

The Big Idea Sources of energy differ in quantity, distribution, usefulness, and the time required for formation.

This chapter was designed to cover the California Grade 6 Science Standards about energy resources (6.3.b, 6.6.a, and 6.6.b). The chapter also covers the discussion of environmental science and the wise use of resources referred to in Category 1, Criterion 11, in the Criteria for Evaluating Instructional Materials in Science. It follows a chapter that introduced natural resources and then explained material resources. Students will make choices about using resources throughout their lives, so understanding the origin of energy and material resources and the consequences of using these resources is important to all students.

After they have completed this chapter, students will begin a chapter about plate tectonics.

Getting Started

		Teach
	Standards Course of Study	❑ SE **Explore Activity** Spinning in the Wind, p. 157 ❑ **Datasheet B for Explore Activity** Spinning in the Wind ■
Universal Access	Advanced Learners/GATE	❑ **Chapter Starter Transparency** ❑ **Datasheet C for Explore Activity** Spinning in the Wind ■
	Basic Learners	❑ SE **Improving Comprehension,** p. 154 ❑ **Chapter Starter Transparency** ❑ **Datasheet A for Explore Activity** Spinning in the Wind ■
	English Learners	❑ SE **Improving Comprehension,** p. 154 ❑ SE **Unpacking the Standards,** p. 155
	Special Education Students	❑ SE **Improving Comprehension,** p. 154 ❑ SE **Unpacking the Standards,** p. 155
	Struggling Readers	❑ SE **Improving Comprehension,** p. 154

Key

SE Student Edition
TE Teacher's Edition
Chapter Resource File
Workbook

CD or CD-ROM
Transparency

Video
■ Also available in Spanish

All resources listed below are also available on the One-Stop Planner.

The California Science Standards listed below are covered in this chapter:

Focus on Earth Sciences

6.3.b Students know that when fuel is consumed, most of the energy released becomes heat energy.

6.6.a Students know the utility of energy sources is determined by factors that are involved in converting these sources to useful forms and the consequences of the conversion process.

6.6.b Students know different natural energy and material resources, including air, soil, rocks, minerals, petroleum, fresh water, wildlife, and forests, and know how to classify them as renewable or nonrenewable.

Investigation and Experimentation

6.7.a Develop a hypothesis.

6.7.c Construct appropriate graphs from data and develop qualitative statements about the relationships between variables.

Practice	Assess
❑ **SE Organize Activity** Double Door, p. 156	❑ **Chapter Pretest**
❑ **TE Academic Vocabulary,** p. 155	
❑ **TE Using Other Graphic Organizers,** p. 154	

5 Pacing • This section should take approximately 2 days to complete.

Section 1 Fossil Fuels

Key Concept Most of the energy used by humans comes from fossil fuels, which are made up of ancient plant and animal matter that stored energy from the sun.

		Teach
	Standards Course of Study	❑ **Bellringer Transparency** ❑ **PowerPoint® Resources** ❑ SE **Quick Lab** Rock Sponge, p. 160 ❑ **Datasheet B for Quick Lab** Rock Sponge ■ ❑ **E11** Fossil Fuel Reservoirs in Earth ❑ **E12** Formation of Coal
Universal Access	Advanced Learners/GATE	❑ TE **Discussion** Foreign Oil, p. 159 ❑ TE **Wordwise** Root Words, p. 159 ❑ **Datasheet C for Quick Lab** Rock Sponge ■ ❑ **L31 Link to Life Science** A Rock-Layer Sequence
Universal Access	Basic Learners	❑ **Datasheet A for Quick Lab** Rock Sponge ■ ❑ TE **Demonstration** Simulating Reservoirs, p. 161
Universal Access	English Learners	❑ TE **Universal Access** Vocabulary Focus, p. 158 ❑ TE **Wordwise** Root Words, p. 159 ❑ TE **Demonstration** Simulating Reservoirs, p. 161 ❑ TE **Universal Access** Fossil Fuel Vocabulary, p. 161 ❑ TE **Using the Figure** Coal Formation, p. 162
Universal Access	Special Education Students	❑ TE **Demonstration** Simulating Reservoirs, p. 161
Universal Access	Struggling Readers	❑ SE **Reading Strategy** Graphic Organizer, p. 158 ❑ TE **Discussion** Fossil Fuel Use, p. 158 ❑ TE **Universal Access** Using Context Clues, p. 159 ❑ TE **Demonstration** Simulating Reservoirs, p. 161 ❑ TE **Using the Figure** Coal Formation, p. 162

Key

SE Student Edition | TE Teacher's Edition | Chapter Resource File | Workbook | CD or CD-ROM | Transparency | Video | ■ Also available in Spanish

All resources listed below are also available on the One-Stop Planner.

Focus on Earth Sciences: 6.3.b, 6.6.a, 6.6.b
Math: Number Sense 6.2.0
English–Language Arts: Reading 6.2.4; Writing 6.1.0

Practice

- ❑ SE **Section Review,** p. 165
- ❑ **Section Review** ■

- ❑ TE **Debate** Drilling in a Wildlife Refuge, p. 163
- ❑ **SciLinks Activity**

- ❑ TE **Universal Access** Modeling Fossil Fuels, p. 162
- ❑ **Interactive Reader and Study Guide**
- ❑ **Vocabulary and Section Summary A** ■
- ❑ **Reinforcement Worksheet**

- ❑ TE **Debate** Drilling in a Wildlife Refuge, p. 163
- ❑ **Interactive Reader and Study Guide**
- ❑ **Vocabulary and Section Summary A** ■
- ❑ **Vocabulary and Section Summary B**

- ❑ TE **Universal Access** Finding Fossil Fuels, p. 163
- ❑ **Interactive Reader and Study Guide**

- ❑ **Interactive Reader and Study Guide**
- ❑ **Directed Reading A** ■
- ❑ **Directed Reading B**

Assess

- ❑ SE **Standards Checks,** pp. 159, 160, 163, 164
- ❑ TE **Standards Focus,** p. 164
  - Assess
  - Reteach
  - Re-Assess
- ❑ **Section Quiz** ■

- ❑ TE **Connection to Language Arts** Creative Writing, p. 161
- ❑ TE **Connection to Real World** Electrical Energy in Your Community, p. 162

Pacing

• This section should take approximately 2 days to complete.

Section 2 Alternative Energy

Key Concept Each alternative energy resource has both benefits and drawbacks.

		Teach
	Standards Course of Study	❑ **Bellringer Transparency** ❑ **PowerPoint® Resources** ❑ **E13** Generating Energy with Fission ❑ SE **Quick Lab** Solar Collector, p. 169 ❑ **Datasheet B for Quick Lab** Solar Collector ■ ❑ **E14** Using Geothermal Energy
Universal Access	Advanced Learners/GATE	❑ **Datasheet C for Quick Lab** Solar Collector ■ ❑ **Calculator-Based Labs** Solar Homes
	Basic Learners	❑ **Datasheet A for Quick Lab** Solar Collector ■ ❑ TE **Reading Strategy** Mnemonics, p. 167 ❑ TE **Reading Strategy** Prediction Guide, p. 170
	English Learners	❑ TE **Wordwise** Word Parts, p. 170
	Special Education Students	
	Struggling Readers	❑ SE **Reading Strategy** Outlining, p. 166 ❑ TE **Discussion** Nuclear Energy, p. 166 ❑ TE **Reading Strategy** Mnemonics, p. 167 ❑ TE **Universal Access** Finding Fossil Fuels, p. 163 ❑ TE **Reading Strategy** Prediction Guide, p. 170 ❑ TE **Wordwise** Word Parts, p. 170

Key

- **SE** Student Edition
- **TE** Teacher's Edition
- Chapter Resource File
- Workbook
- CD or CD-ROM
- Transparency
- Video
- ■ Also available in Spanish

All resources listed below are also available on the One-Stop Planner.

Focus on Earth Sciences: 6.3.b, 6.6.a, 6.6.b
Math: Number Sense 6.2.0

Practice	Assess
❑ **SE Section Review,** p. 173 ❑ **Section Review** ■	❑ **SE Standards Checks,** pp. 167, 168, 169, 170 ❑ **TE Standards Focus,** p. 172 • Assess • Reteach • Re-Assess ❑ **Section Quiz** ■
❑ **TE Connection to History** History of Nuclear Energy, p. 168 ❑ **TE Connection to Astronomy** Conserving Energy in Space, p. 170 ❑ **TE Activity** Burning Biomass Fuel, p. 171 ❑ **Critical Thinking** ❑ **Long-Term Projects & Research Ideas** Canyon Controversy ❑ **Long-Term Projects & Research Ideas** To Complicate Things	❑ **TE Universal Access** Comparing Nuclear and Coal-Fired Power Plants, p. 171 ❑ **TE Homework** The Aftermath of Chernobyl, p. 167 ❑ **Long-Term Projects & Research Ideas** Meltdown! ❑ **Long-Term Projects & Research Ideas** Build a City—Save a World!
❑ **Interactive Reader and Study Guide** ❑ **Vocabulary and Section Summary A** ■	
❑ **TE Using the Figure** Wind Energy, p. 168 ❑ **Interactive Reader and Study Guide** ❑ **Vocabulary and Section Summary A** ■ ❑ **Vocabulary and Section Summary B**	
❑ **TE Universal Access** Fission Pros and Cons, p. 167 ❑ **TE Universal Access** Alternative Energies Outline, p. 172 ❑ **Interactive Reader and Study Guide**	
❑ **TE Using the Figure** Wind Energy, p. 168 ❑ **TE Universal Access** Comparing and Contrasting, p. 169 ❑ **Interactive Reader and Study Guide** ❑ **Directed Reading A** ■ ❑ **Directed Reading B**	❑ **TE Universal Access** Activating Prior Knowledge, p. 168

5 Pacing

- Chapter Lab, Review, and Assessment should take approximately 3 days to complete.

Wrapping Up

	Teach
Standards Course of Study	❑ SE **Model-Making Lab** Making a Water Wheel, pp. 174–175 ❑ **Datasheet B for Chapter Lab** Making a Water Wheel ■ ❑ **Standards Review Transparency** ■
Universal Access: Advanced Learners/GATE	❑ **Datasheet C for Chapter Lab** Making a Water Wheel ■
Universal Access: Basic Learners	❑ **Datasheet A for Chapter Lab** Making a Water Wheel ■
Universal Access: English Learners	
Universal Access: Special Education Students	
Universal Access: Struggling Readers	

Additional Resources

SUPER SUMMARY

Have students review the major concepts in this chapter by using the Super Summary that includes the following:

- an outline of important points in the chapter
- flashcards for chapter vocabulary
- an interactive quiz

Go to **go.hrw.com**
Type in the keyword HY7ENRS

Performance-Based Assessments

The Chapter Resource File for this chapter contains a hands-on activity that can be used to help assess student progress in a nontraditional format. In the Performance-Based Assessment for this chapter, students study the effects of an acid solution that is similar to acid precipitation on two types of rock.

Key

SE Student Edition
TE Teacher's Edition
Chapter Resource File
Workbook
CD or CD-ROM
Transparency
Video
■ Also available in Spanish

All resources listed below are also available on the One-Stop Planner.

Focus on Earth Sciences: 6.3.b, 6.6.a, 6.6.b, 6.7.a, 6.7.c
Math: Number Sense 6.2.0
English–Language Arts: Writing 6.1.0

Practice	Assess
❑ SE **Science Skills Activity** Constructing Graphs from Data, p. 176 ❑ **Datasheet for Science Skills Activity** ■ ❑ **Concept Mapping Transparency** ❑ SE **Chapter Review,** pp. 178–179 ❑ **Chapter Review** ■	❑ **Chapter Test B** ■ ❑ SE **Standards Assessment,** pp. 180–181 ❑ **Standards Assessment** ❑ **Standards Review Workbook** ■
	❑ SE **Language Arts Activity,** p. 182 ❑ SE **Math Activity,** p. 182 ❑ SE **Social Studies Activity,** p. 183 ❑ **Chapter Test C** ❑ **Brain Food Video Quiz**
❑ TE **Focus on Speaking** Explaining the Standards, p. 177	❑ **Chapter Test A** ❑ **Brain Food Video Quiz**
❑ TE **Identifying Prefixes** Defining Prefixes, p. 177 ❑ TE **Focus on Speaking** Explaining the Standards, p. 177	❑ **Brain Food Video Quiz**
❑ TE **Universal Access** Group Discussion, p. 176 ❑ TE **Focus on Speaking** Explaining the Standards, p. 177	❑ **Brain Food Video Quiz**
❑ TE **Universal Access** Following Directions, p. 176	❑ **Brain Food Video Quiz**

Holt Online Assessment

Post tests and quizzes to Holt Online Assessment, an assessment management tool. The system automatically grades the assessments, and you receive students' scores and information about which questions students missed. Holt Online Assessment is available through the Premier Online Edition of *Holt California Earth Science.*

Holt Anthology of Science Fiction

The Holt Anthology of Science Fiction includes thought-provoking stories that are relevant to science instruction. Enhance students' learning by asking them to read a story from the *Holt Anthology of Science Fiction* and to answer questions about what they have read.

6

Pacing

- This chapter should take approximately 11 days to complete.
- Getting Started should take approximately 1 day to complete.

Chapter 6 Plate Tectonics

The Big Idea Plate tectonics accounts for important features of Earth's surface and major geologic events.

This chapter was designed to cover the California Grade 6 Science Standards about plate tectonics (6.1.a, 6.1.b, 6.1.c, 6.1.d, 6.1.e, 6.1.f, and 6.4.c). In this chapter, the theory of plate tectonics is presented and the relationship between plate tectonics and Earth's surface features is explored.

After they have completed this chapter, students will begin a chapter about earthquakes and will learn how earthquakes are related to plate tectonics.

Getting Started

		Teach
	Standards Course of Study	❑ **SE Explore Activity** Continental Collisions, p. 189 ❑ **Datasheet B for Explore Activity** Continental Collisions ■
Universal Access	Advanced Learners/GATE	❑ **Chapter Starter Transparency** ❑ **Datasheet C for Explore Activity** Continental Collisions ■
Universal Access	Basic Learners	❑ **SE Improving Comprehension,** p. 186 ❑ **Chapter Starter Transparency** ❑ **Datasheet A for Explore Activity** Continental Collisions ■
Universal Access	English Learners	❑ **SE Improving Comprehension,** p. 186 ❑ **SE Unpacking the Standards,** p. 187
Universal Access	Special Education Students	❑ **SE Improving Comprehension,** p. 186 ❑ **SE Unpacking the Standards,** p. 187
Universal Access	Struggling Readers	❑ **SE Improving Comprehension,** p. 186

Key

SE Student Edition
TE Teacher's Edition
Chapter Resource File
Workbook
CD or CD-ROM
Transparency

Video
■ Also available in Spanish

All resources listed below are also available on the One-Stop Planner.

The California Science Standards listed below are covered in this chapter:

Focus on Earth Sciences

6.1.a Students know evidence of plate tectonics is derived from the fit of the continents; the location of earthquakes, volcanoes, and midocean ridges; and the distribution of fossils, rock types, and ancient climatic zones.

6.1.b Students know Earth is composed of several layers: a cold, brittle lithosphere; a hot, convecting mantle; and a dense, metallic core.

6.1.c Students know lithospheric plates the size of continents and oceans move at rates of centimeters per year in response to movements in the mantle.

6.1.d Students know that earthquakes are sudden motions along breaks in the crust called faults and that volcanoes and fissures are locations where magma reaches the surface.

6.1.e Students know major geologic events, such as earthquakes, volcanic eruptions, and mountain building, result from plate motions.

6.1.f Students know how to explain major features of California geology (including mountains, faults, volcanoes) in terms of plate tectonics.

6.4.c Students know heat from Earth's interior reaches the surface primarily through convection.

Investigation and Experimentation

6.7.g Interpret events by sequence and time from natural phenomena (e.g., the relative ages of rocks and intrusions).

Practice	Assess
❑ **SE Organize Activity** Key-Term Fold, p. 188	❑ **Chapter Pretest**
❑ TE **Words with Multiple Meanings,** p. 187	
❑ TE **Using Other Graphic Organizers,** p. 186	

6

Pacing

• This section should take approximately 1.5 days to complete.

Section 1 Earth's Structure

Key Concept Earth is composed of several layers. The continents are part of the uppermost layer, and they move slowly around Earth's surface.

	Teach
Standards Course of Study	❑ **Bellringer Transparency** ❑ **PowerPoint® Resources** ❑ SE **Quick Lab** Making Magnets, p. 195 ❑ **Datasheet B for Quick Lab** Making Magnets ■ ❑ **E15** The Composition of Earth and the Physical Structure of Earth ❑ **E16** Discoveries of Earth's Interior ❑ **E17** Sea-Floor Spreading
Universal Access — **Advanced Learners/GATE**	❑ **Datasheet C for Quick Lab** Making Magnets ■
Universal Access — **Basic Learners**	❑ TE **Using the Figure** Earth's Layers, p. 190 ❑ TE **Connection to Physical Science** Waves and Energy, p. 192 ❑ TE **Connection to Physical Science** Sonar, p. 194 ❑ TE **Using the Figure** Magnetic Anomalies, p. 195 ❑ **Datasheet A for Quick Lab** Making Magnets ■
Universal Access — **English Learners**	❑ TE **Wordwise** Identifying Roots, p. 191
Universal Access — **Special Education Students**	❑ TE **Universal Access** Avocado Earth, p. 190 ❑ TE **Using the Figure** Two Earth Models, p. 191 ❑ TE **Using the Figure** Evidence for Continental Drift, p. 193
Universal Access — **Struggling Readers**	❑ SE **Reading Strategy** Graphic Organizer, p. 190

Key

SE Student Edition | TE Teacher's Edition | Chapter Resource File | Workbook | CD or CD-ROM | Transparency | Video | ■ Also available in Spanish

All resources listed below are also available on the One-Stop Planner.

Focus on Earth Sciences: 6.1.a, 6.1.b
Math: Algebra and Functions 6.2.3
English–Language Arts: Reading 6.2.4

Practice	Assess
❑ SE **Section Review,** p. 197 ❑ **Section Review** ■	❑ SE **Standards Checks,** pp. 190, 192, 193, 194 ❑ TE **Standards Focus,** p. 196 • Assess • Reteach • Re-Assess ❑ **Section Quiz** ■
❑ TE **Universal Access** Planet Composition Comparison, p. 190	❑ TE **Connection to Life Science** The Breakup of Pangaea and Dinosaur Evolution, p. 194
❑ TE **Universal Access** Modeling Layers, p. 191 ❑ TE **Group Activity** Map Puzzle, p. 193 ❑ TE **Universal Access** Modeling Sea-Floor Spreading, p. 196 ❑ **Interactive Reader and Study Guide** ❑ **Vocabulary and Section Summary A** ■ ❑ **Reinforcement Worksheet** ❑ **Labs You Can Eat** Rescue Near the Center of the Earth	
❑ **Interactive Reader and Study Guide** ❑ **Vocabulary and Section Summary A** ■ ❑ **Vocabulary and Section Summary B**	
❑ **Interactive Reader and Study Guide**	
❑ TE **Universal Access** Scanning for Key Words, p. 192 ❑ **Interactive Reader and Study Guide** ❑ **Directed Reading A** ■ ❑ **Directed Reading B**	❑ TE **Universal Access** Evaluating Evidence, p. 193

6 **Pacing** • This section should take approximately 1.5 days to complete.

Section 2 The Theory of Plate Tectonics

Key Concept Tectonic plates the size of continents and oceans move at rates of a few centimeters per year in response to movements in the mantle.

	Teach
Standards Course of Study	❑ **Bellringer Transparency** ❑ **PowerPoint® Resources** ❑ SE **Quick Lab** Tectonic Ice Cubes, p. 200 ❑ **Datasheet B for Quick Lab** Tectonic Ice Cubes ■ ❑ **E18** Earth's Major Tectonic Plates and A Tectonic Plate Close-Up ❑ **E19** Tectonic Plate Boundaries ❑ **E20** Three Possible Driving Forces of Plate Tectonics
Universal Access: Advanced Learners/GATE	❑ TE **Universal Access** A Dynamic Tectonic Plate Map, p. 198 ❑ **Datasheet C for Quick Lab** Tectonic Ice Cubes ■
Universal Access: Basic Learners	❑ TE **Discussion** Sea-Floor Spreading, p. 198 ❑ **Datasheet A for Quick Lab** Tectonic Ice Cubes ■
Universal Access: English Learners	❑ TE **Discussion** A Preposterous Theory, p. 199
Universal Access: Special Education Students	❑ TE **Using the Figure** Plate Boundaries, p. 201
Universal Access: Struggling Readers	❑ SE **Reading Strategy** Clarifying Concepts, p. 198 ❑ TE **Reading Strategy** Reading Organizer, p. 199 ❑ TE **Universal Access** Identifying Structural Patterns, p. 202

Key

SE Student Edition
TE Teacher's Edition
Chapter Resource File
Workbook
CD or CD-ROM
Transparency
Video
■ Also available in Spanish

All resources listed below are also available on the One-Stop Planner.

Focus on Earth Sciences: 6.1.b, 6.1.c, 6.1.e, 6.4.c
Math: Algebra and Functions 6.2.3
English–Language Arts : Reading 6.1.1

Practice	Assess
❑ SE **Section Review,** p. 203 ❑ **Section Review** ■	❑ SE **Standards Checks,** pp. 199, 201, 202, 203 ❑ TE **Standards Focus,** p. 202 • Assess • Reteach • Re-Assess ❑ **Section Quiz** ■
❑ **SciLinks Activity**	
❑ TE **Universal Access** Tectonic Plate Boundary Cause-and-Effect Chart, p. 200 ❑ TE **Group Activity** Modeling Tectonic Plates, p. 201 ❑ **Interactive Reader and Study Guide** ❑ **Vocabulary and Section Summary A** ■	❑ TE **Connection to Math** Computation, p. 200
❑ TE **Universal Access** Illustrating Plate Movements, p. 201 ❑ **Interactive Reader and Study Guide** ❑ **Vocabulary and Section Summary A** ■ ❑ **Vocabulary and Section Summary B**	
❑ **Interactive Reader and Study Guide** ❑ **Labs You Can Eat** Cracks in the Hard-Boiled Earth	❑ TE **Universal Access** Tectonic Art, p. 201
❑ **Interactive Reader and Study Guide** ❑ **Directed Reading A** ■ ❑ **Directed Reading B**	

6 Pacing • This section should take approximately 1.5 days to complete.

Section 3 Deforming Earth's Crust

Key Concept Tectonic plate motions deform Earth's crust. Deformation causes rock layers to bend and break and causes mountains to form.

	Teach
Standards Course of Study	❑ **Bellringer Transparency** ❑ **PowerPoint® Resources** ❑ SE **Quick Lab** Modeling Strike-Slip Faults, p. 207 ❑ **Datasheet B for Quick Lab** Modeling Strike-Slip Faults ■ ❑ **E21** Normal, Reverse, and Strike-Slip Faults
Universal Access: Advanced Learners/GATE	❑ TE **Discussion** Folds That Trap Natural Gas, p. 205 ❑ **Datasheet C for Quick Lab** Modeling Strike-Slip Faults ■
Universal Access: Basic Learners	❑ TE **Demonstration** Modeling Deformation, p. 204 ❑ TE **Using the Figure** Direction and Parts of Folds, p. 205 ❑ TE **Using the Figure** Hanging Walls Versus Footwalls, p. 206 ❑ **Datasheet A for Quick Lab** Modeling Strike-Slip Faults ■
Universal Access: English Learners	
Universal Access: Special Education Students	❑ TE **Universal Access** Symmetrical and Asymmetrical, p. 200
Universal Access: Struggling Readers	❑ SE **Reading Strategy** Summarizing, p. 204 ❑ TE **Universal Access** Using Text Features, p. 206 ❑ TE **Universal Access** Previewing, p. 208

Additional Resources

Holt Lab Generator CD-ROM

Search for any lab by topic, standard, difficulty level, or time. Edit any lab to fit your needs, or create your own labs. Use the Lab Materials QuickList software to customize your lab materials list. Lab datasheets are also available in Spanish on this CD-ROM.

Guided Reading Audio CD Program

The Guided Reading Audio CD Program provides a direct reading of the student text. This resource is helpful to auditory learners and struggling readers. This program is available in English and Spanish.

Key

SE Student Edition
TE Teacher's Edition
Chapter Resource File
Workbook
CD or CD-ROM
Transparency
Video
■ Also available in Spanish

All resources listed below are also available on the One-Stop Planner.

Focus on Earth Sciences: 6.1.a, 6.1.d, 6.1.e

Practice	Assess
❑ SE **Section Review,** p. 209 ❑ **Section Review** ■	❑ SE **Standards Checks,** pp. 204, 206, 207, 208 ❑ TE **Standards Focus,** p. 208 • Assess • Reteach • Re-Assess ❑ **Section Quiz** ■
❑ **Labs You Can Eat** Dough Fault of Your Own	❑ TE **Universal Access** Short Story, p. 204
❑ TE **Homework** Making Models, p. 200 ❑ TE **Universal Access** Mountain Models, p. 205 ❑ **Interactive Reader and Study Guide** ❑ **Vocabulary and Section Summary A** ■	
❑ TE **Universal Access** Local Topography, p. 207 ❑ **Interactive Reader and Study Guide** ❑ **Vocabulary and Section Summary A** ■ ❑ **Vocabulary and Section Summary B**	
❑ **Interactive Reader and Study Guide**	❑ TE **Activity** Modeling Fault Motion, p. 207
❑ TE **Reading Strategy** Reading Organizer, p. 206 ❑ **Interactive Reader and Study Guide** ❑ **Directed Reading A** ■ ❑ **Directed Reading B** ■	

Reviewing Prior Knowledge

Prepare students to learn about the deformation of Earth's surface by summarizing standard 6.1.e. See page 204 of the Teacher's Edition.

Bendable Rock Students may have difficulty believing that it is possible for massive blocks of rock to bend. To correct this misconception, see page 206 in the Teacher's Edition.

Section 4 California Geology

Key Concept Major features of California geology can be explained in terms of plate tectonics.

		Teach
	Standards Course of Study	❑ **Bellringer Transparency** ❑ **PowerPoint® Resources** ❑ SE **Quick Lab** Modeling Accretion, p. 213 ❑ **Datasheet B for Quick Lab** Modeling Accretion ■ ❑ **E22** Changing Tectonic Plate Boundaries ❑ **E23** Geologic Map of California
Universal Access	Advanced Learners/GATE	❑ **L36 Link to Life Science** Formation of the Panama Land Bridge ❑ **Datasheet C for Quick Lab** Modeling Accretion ■
Universal Access	Basic Learners	❑ TE **Universal Access** Finding the Fault, p. 214 ❑ **Datasheet A for Quick Lab** Modeling Accretion ■
Universal Access	English Learners	❑ TE **Universal Access** Words with Multiple Meanings, p. 211 ❑ TE **Wordwise** Defining Vocabulary, p. 212
Universal Access	Special Education Students	
Universal Access	Struggling Readers	❑ SE **Reading Strategy** Graphic Organizer, p. 210 ❑ TE **Universal Access** Identifying Structural Patterns, p. 210 ❑ TE **Reading Strategy** Paired Summarizing, p. 212

Additional Resources

Holt Lab Generator CD-ROM

Search for any lab by topic, standard, difficulty level, or time. Edit any lab to fit your needs, or create your own labs. Use the Lab Materials QuickList software to customize your lab materials list. Lab datasheets are also available in Spanish on this CD-ROM.

Guided Reading Audio CD Program

The Guided Reading Audio CD Program provides a direct reading of the student text. This resource is helpful to auditory learners and struggling readers. This program is available in English and Spanish.

Key

SE Student Edition	Chapter Resource File	CD or CD-ROM	Video
TE Teacher's Edition	Workbook	Transparency	■ Also available in Spanish

All resources listed below are also available on the One-Stop Planner.

Focus on Earth Sciences: 6.1.f
Math: Number Sense 6.2.1
English–Language Arts: Reading 6.2.4

Practice	Assess
❑ **SE Section Review,** p. 217 ❑ **Section Review** ■	❑ **SE Standards Checks,** pp. 210, 213, 214, 216 ❑ **TE Standards Focus,** p. 216 • Assess • Reteach • Re-Assess ❑ **Section Quiz** ■
❑ **TE Activity** Collage Project, p. 210 ❑ **TE Universal Access** Ballad of the California Landscape, p. 213 ❑ **TE Cultural Awareness** Research California Diversity and the Gold Rush, p. 213 ❑ **Critical Thinking**	❑ **TE Connection to Astronomy** Extraterrestrial Mountains, p. 214
❑ **TE Universal Access** Offset Measurement, p. 212 ❑ **TE Activity** Lesson Plan on California Geology, p. 215 ❑ **Interactive Reader and Study Guide** ❑ **Vocabulary and Section Summary A** ■	❑ **TE Using the Figure** Changing Tectonic Plate Boundaries, p. 211 ❑ **TE Homework** A Tectonic History of California, p. 211
❑ **Interactive Reader and Study Guide** ❑ **Vocabulary and Section Summary A** ■ ❑ **Vocabulary and Section Summary B**	
❑ **TE Universal Access** Modeling Fault Systems, p. 215 ❑ **Interactive Reader and Study Guide**	
❑ **Interactive Reader and Study Guide** ❑ **Directed Reading A** ■ ❑ **Directed Reading B**	❑ **TE Activity** The "Solid" Earth, p. 214

Reviewing Prior Knowledge

Prepare students to learn about the geology of California by identifying the tectonic plate boundaries that impact California geology. See page 210 of the Teacher's Edition.

San Andreas Fault Students may think that the San Andreas fault is a single fault that divides the North American and Pacific plates. To correct this misconception, see page 215 in the Teacher's Edition.

Math Support

Science and math go hand in hand. The Math Practice on page 216 helps students practice math skills in a scientific context.

Pacing

- Chapter Lab, Review, and Assessment should take approximately 3 days to complete.

Wrapping Up

		Teach
	Standards Course of Study	❑ SE **Model-Making Lab** Sea-Floor Spreading, pp. 218–219 ❑ **Datasheet B for Chapter Lab** Sea-Floor Spreading ■ ❑ **Standards Review Transparency** ■
Universal Access	Advanced Learners/GATE	❑ **Datasheet C for Chapter Lab** Sea-Floor Spreading ■
Universal Access	Basic Learners	❑ **Datasheet A for Chapter Lab** Sea-Floor Spreading ■
Universal Access	English Learners	❑ TE **Identifying Roots** Vocabulary Word Parts, p. 221
Universal Access	Special Education Students	
Universal Access	Struggling Readers	❑ TE **Focus on Reading** Comparing Texts, p. 221

Additional Resources

SUPER SUMMARY

Have students review the major concepts in this chapter by using the Super Summary that includes the following:

- an outline of important points in the chapter
- flashcards for chapter vocabulary
- an interactive quiz

Go to **go.hrw.com**
Type in the keyword HY7TECS

Performance-Based Assessments

The Chapter Resource File for this chapter contains a hands-on activity that can be used to help assess student progress in a nontraditional format. In the Performance-Based Assessment for this chapter, students use maps to find a fault line.

Key

SE Student Edition
TE Teacher's Edition

Chapter Resource File
Workbook

CD or CD-ROM
Transparency

Video
■ Also available in Spanish

All resources listed below are also available on the One-Stop Planner.

Focus on Earth Sciences: 6.1.a, 6.1.b, 6.1.c, 6.1.d, 6.1.e, 6.1.f, 6.4.c, 6.7.g
Math: Algebra and Functions 6.2.3
English–Language Arts: Writing 6.1.3

Practice	Assess
❑ SE **Science Skills Activity** Interpreting Time from Natural Phenomena, p. 220 ❑ **Datasheet for Science Skills Activity** ■ ❑ **Concept Mapping Transparency** ❑ SE **Chapter Review,** pp. 222–223 ❑ **Chapter Review** ■	❑ **Chapter Test B** ■ ❑ SE **Standards Assessment,** pp. 224–225 ❑ **Standards Assessment** ❑ **Standards Review Workbook** ■
	❑ **Chapter Test C** ■ ❑ **Brain Food Video Quiz**
	❑ TE **Social Studies Activity,** p. 226 ❑ TE **Math Activity,** p. 227 ❑ **Chapter Test A** ❑ **Brain Food Video Quiz**
	❑ TE **Language Arts Activity,** p. 226 ❑ **Brain Food Video Quiz**
❑ TE **Universal Access** Understanding Polarity, p. 220	❑ **Brain Food Video Quiz**
	❑ **Brain Food Video Quiz**

Holt Online Assessment

Post tests and quizzes to Holt Online Assessment, an assessment management tool. The system automatically grades the assessments, and you receive students' scores and information about which questions students missed. Holt Online Assessment is available through the Premier Online Edition of *Holt California Earth Science.*

Holt Anthology of Science Fiction

The Holt Anthology of Science Fiction includes thought-provoking stories that are relevant to science instruction. Enhance students' learning by asking them to read a story from the *Holt Anthology of Science Fiction* and to answer questions about what they have read.

7 **Pacing**
- **This chapter should take approximately 11 days to complete.**
- **Getting Started should take approximately 1 day to complete.**

Chapter 7 Earthquakes

The Big Idea Earthquakes result from sudden motions along breaks in Earth's crust and can affect landforms and societies.

This chapter was designed to cover the California Grade 6 Science Standards about earthquakes (6.1.a, 6.1.d, 6.1.e, 6.1.g, 6.2.d, and 6.3.a). It follows a chapter about plate tectonics because understanding the movement of tectonic plates provides a base on which students can build an understanding of earthquakes.

After they have completed this chapter, students will begin a chapter about volcanoes and the relationship between volcanoes and plate tectonics.

Getting Started

		Teach
	Standards Course of Study	❑ SE **Explore Activity** Investigating Building Materials, p. 231 ❑ **Datasheet B for Explore Activity** Investigating Building Materials ■
Universal Access	Advanced Learners/GATE	❑ **Chapter Starter Transparency** ❑ **Datasheet C for Explore Activity** Investigating Building Materials ■
Universal Access	Basic Learners	❑ SE **Improving Comprehension,** p. 228 ❑ **Chapter Starter Transparency** ❑ **Datasheet A for Explore Activity** Investigating Building Materials ■
Universal Access	English Learners	❑ SE **Improving Comprehension,** p. 228 ❑ SE **Unpacking the Standards,** p. 229
Universal Access	Special Education Students	❑ SE **Improving Comprehension,** p. 228 ❑ SE **Unpacking the Standards,** p. 229
Universal Access	Struggling Readers	❑ SE **Improving Comprehension,** p. 228

Key

SE Student Edition
TE Teacher's Edition
Chapter Resource File
Workbook

CD or CD-ROM
Transparency

Video
■ Also available in Spanish

All resources listed below are also available on the One-Stop Planner.

The California Science Standards listed below are covered in this chapter:

Focus on Earth Sciences

6.1.a Students know evidence of plate tectonics is derived from the fit of the continents; the location of earthquakes, volcanoes, and midocean ridges; and the distribution of fossils, rock types, and ancient climatic zones.

6.1.d Students know that earthquakes are sudden motions along breaks in the crust called faults and that volcanoes and fissures are locations where magma reaches the surface.

6.1.e Students know major geologic events, such as earthquakes, volcanic eruptions, and mountain building, result from plate motions.

6.1.g Students know how to determine the epicenter of an earthquake and know that the effects of an earthquake on any region vary, depending on the size of the earthquake, the distance of the region from the epicenter, the local geology, and the type of construction in the region.

6.2.d Students know earthquakes, volcanic eruptions, landslides, and floods change human and wildlife habitats.

6.3.a Students know energy can be carried from one place to another by heat flow or by waves, including water, light and sound waves, or by moving objects.

Investigation and Experimentation

6.7.b Select and use appropriate tools and technology (including calculators, computers, balances, spring scales, microscopes, and binoculars) to perform tests, collect data, and display data.

6.7.c Construct appropriate graphs from data and develop qualitative statements about the relationships between variables.

Practice	Assess
❑ SE **Organize Activity** Pyramid, p. 230	❑ **Chapter Pretest**
❑ TE **Word with Multiple Meanings,** p. 229	
❑ TE **Using Other Graphic Organizers,** p. 228	

Pacing

• This section should take approximately 2.5 days to complete.

Section 1 What Are Earthquakes?

Key Concept Sudden motions along breaks in Earth's crust can release energy in the form of seismic waves.

	Teach
Standards Course of Study	❑ **Bellringer Transparency** ❑ **PowerPoint® Resources** ❑ SE **Quick Lab** Seismic Spring Toys, p. 236 ❑ **Datasheet B for Quick Lab** Seismic Spring Toys ■ ❑ **E24** Locations of Earthquakes ❑ **E25** Earthquakes at Divergent Boundaries ❑ **E26** Earthquakes at Convergent Boundaries ❑ **E27** Earthquakes at Transform Boundaries ❑ **E28** Elastic Rebound ❑ **E29** Types of Seismic Waves
Universal Access	
Advanced Learners/GATE	❑ **Datasheet C for Quick Lab** Seismic Spring Toys ■
Basic Learners	❑ **Datasheet A for Quick Lab** Seismic Spring Toys ■
English Learners	❑ TE **Discussion** Seismic Definitions, p. 232
Special Education Students	❑ TE **Demonstration** Faults and Earthquakes, p. 234
Struggling Readers	❑ SE **Reading Strategy** Graphic Organizer, p. 232 ❑ TE **Universal Access** Words with Multiple Meanings, p. 232 ❑ TE **Reading Strategy** Plate Boundaries, p. 233

Additional Resources

Holt Lab Generator CD-ROM

Search for any lab by topic, standard, difficulty level, or time. Edit any lab to fit your needs, or create your own labs. Use the Lab Materials QuickList software to customize your lab materials list. Lab datasheets are also available in Spanish on this CD-ROM.

Guided Reading Audio CD Program

The Guided Reading Audio CD Program provides a direct reading of the student text. This resource is helpful to auditory learners and struggling readers. This program is available in English and Spanish.

Key

SE Student Edition
TE Teacher's Edition

Chapter Resource File
Workbook

CD or CD-ROM
Transparency

Video
■ Also available in Spanish

All resources listed below are also available on the One-Stop Planner.

Focus on Earth Sciences: 6.1.a, 6.1.d, 6.1.e, 6.3.a
English–Language Arts: Reading 6.2.4

Practice	Assess
❑ SE **Section Review,** p. 237 ❑ **Section Review** ■	❑ SE **Standards Checks,** pp. 232, 234, 235, 236, 237 ❑ TE **Standards Focus,** p. 236 • Assess • Reteach • Re-Assess ❑ **Section Quiz** ■
❑ TE **Universal Access** Mapping Earthquakes, p. 234 ❑ **SciLinks Activity**	
❑ TE **Universal Access** Modeling Deformation, p. 233 ❑ **Interactive Reader and Study Guide** ❑ **Vocabulary and Section Summary A** ■	
❑ TE **Universal Access** Demonstrating Earthquakes, p. 234 ❑ **Interactive Reader and Study Guide** ❑ **Vocabulary and Section Summary A** ■ ❑ **Vocabulary and Section Summary B**	❑ TE **Universal Access** Studying Faults, p. 235
❑ **Interactive Reader and Study Guide**	
❑ **Interactive Reader and Study Guide** ❑ **Directed Reading A** ■ ❑ **Directed Reading B**	

Reviewing Prior Knowledge

Prepare students to learn about how earthquakes are related to tectonic plate boundaries and review the locations of tectonic plate boundaries. See page 232 of the Teacher's Edition.

MISCONCEPTION ALERT

Aftershocks Students may think that aftershocks do not present the same danger as the mainshock. To correct this misconception, see page 235 in the Teacher's Edition.

Pacing • This section should take approximately 2.5 days to complete.

Section 2 Earthquake Measurement

Key Concept Studying the properties of seismic waves can help scientists determine an earthquake's starting point, strength, and intensity.

	Teach
Standards Course of Study	❑ **Bellringer Transparency** ❑ **PowerPoint® Resources** ❑ **SE Quick Lab** Locating an Epicenter, p. 239 ❑ **Datasheet B for Quick Lab** Locating an Epicenter ■ ❑ **E30** Finding an Earthquake's Epicenter ❑ **E31** Fault Map of California and Locations of Faults in California ❑ **E32** Modified Mercalli Scale
Universal Access: Advanced Learners/GATE	❑ **Datasheet C for Quick Lab** Locating an Epicenter ■
Universal Access: Basic Learners	❑ **Datasheet A for Quick Lab** Locating an Epicenter ■
Universal Access: English Learners	
Universal Access: Special Education Students	
Universal Access: Struggling Readers	❑ **SE Reading Strategy** Summarizing, p. 238

Key

SE Student Edition
TE Teacher's Edition
Chapter Resource File
Workbook
CD or CD-ROM
Transparency
Video
■ Also available in Spanish

All resources listed below are also available on the One-Stop Planner.

Focus on Earth Sciences: 6.1.g
Math: Algebra and Functions 6.2.3; Mathematical Reasoning 6.2.4

Practice	Assess
❑ SE **Section Review,** p. 243 ❑ **Section Review** ■	❑ SE **Standards Checks,** pp. 240, 241, 243 ❑ TE **Standards Focus,** p. 242 • Assess • Reteach • Re-Assess ❑ **Section Quiz** ■
❑ TE **Activity** Poster Project, p. 238	
❑ TE **Universal Access** Traveling Waves, p. 238 ❑ TE **Using the Figure** Locating Earthquakes, p. 240 ❑ TE **Connection to Geography** Location History, p. 240 ❑ TE **Using the Figure** Mercalli Intensity Scale, p. 241 ❑ **Interactive Reader and Study Guide** ❑ **Vocabulary and Section Summary A** ■ ❑ **Reinforcement Worksheet**	❑ TE **Using the Figure** Finding an Epicenter, p. 239
❑ **Interactive Reader and Study Guide** ❑ **Vocabulary and Section Summary A** ■ ❑ **Vocabulary and Section Summary B**	❑ TE **Universal Access** Earthquakes in the News, p. 241 ❑ TE **Universal Access** Nightly News, p. 242
❑ TE **Universal Access** Earthquake Effects, p. 239 ❑ TE **Universal Access** Modeling Intensity, p. 241 ❑ **Interactive Reader and Study Guide**	
❑ **Interactive Reader and Study Guide** ❑ **Directed Reading A** ■ ❑ **Directed Reading B**	❑ TE **Universal Access** Key Word Cards, p. 242

7 **Pacing** • This section should take approximately 2.5 days to complete.

Section 3 Earthquakes and Society

Key Concept Studying seismic activity can help scientists forecast earthquakes and reduce the damage that earthquakes cause.

		Teach
	Standards Course of Study	❑ **Bellringer Transparency** ❑ **PowerPoint® Resources** ❑ **SE Quick Lab** Earthquakes and Buildings, p. 246 ❑ **Datasheet B for Quick Lab** Earthquakes and Buildings ■ ❑ **SE Quick Lab** Modeling a Tsunami, p. 250 ❑ **Datasheet B for Quick Lab** Modeling a Tsunami ■ ❑ **E33** Earthquake Hazard Level Map ❑ **E34** Earthquake-Resistant Building Technology ❑ **E35** Formation of Tsunamis
Universal Access	Advanced Learners/GATE	❑ **L29 Link to Life Science** Hutton and the Principle of Uniformitarianism ❑ **Datasheet C for Quick Lab** Earthquakes and Buildings ■ ❑ **Datasheet C for Quick Lab** Modeling a Tsunami ■
Universal Access	Basic Learners	❑ **TE Using the Figure** Earthquake Resistant Building Technology, p. 244 ❑ **TE Discussion** Hazard Levels, p. 244 ❑ **Datasheet A for Quick Lab** Earthquakes and Buildings ■ ❑ **Datasheet A for Quick Lab** Modeling a Tsunami ■
Universal Access	English Learners	❑ **TE Universal Access** Summarizing Forecasts, p. 244
Universal Access	Special Education Students	
Universal Access	Struggling Readers	❑ **SE Reading Strategy** Graphic Organizer, p. 244 ❑ **TE Reading Strategy** Prediction Guide, p. 245 ❑ **TE Universal Access** Evaluating Evidence, p. 246

Key

SE Student Edition · TE Teacher's Edition · Chapter Resource File · Workbook · CD or CD-ROM · Transparency · Video · ■ Also available in Spanish

All resources listed below are also available on the One-Stop Planner.

Focus on Earth Sciences: 6.1.g, 6.2.d
Math: Algebra and Functions 6.2.3; Mathematical Reasoning 6.2.4
English–Language Arts: Reading 6.2.4

Practice	Assess
❑ SE **Section Review,** p. 251 ❑ **Section Review** ■	❑ SE **Standards Checks,** pp. 245, 246, 248, 249, 250 ❑ TE **Standards Focus,** p. 250 • Assess • Reteach • Re-Assess ❑ **Section Quiz** ■
❑ TE **Universal Access** Debate, p. 245 ❑ TE **Activity** Poster Project, p. 249 ❑ **Critical Thinking**	❑ **Long-Term Projects & Research Ideas** A Whole Lotta Shakin'
❑ TE **Universal Access** Quiz Game, p. 244 ❑ TE **Group Activity** Earthquake Safety, p. 248 ❑ TE **Demonstration** Giant Wave, p. 248 ❑ **Interactive Reader and Study Guide** ❑ **Vocabulary and Section Summary A** ■	❑ TE **Homework** Presentation, p. 246 ❑ TE **Connection to Real Life** Earthquake Kit, p. 249
❑ TE **Universal Access** Earthquakes Around the World, p. 247 ❑ **Interactive Reader and Study Guide** ❑ **Vocabulary and Section Summary A** ■ ❑ **Vocabulary and Section Summary B**	
❑ TE **Universal Access** Earthquake Expectations, p. 245 ❑ TE **Universal Access** Earthquake-Resistant Model, p. 246 ❑ TE **Universal Access** Modeling Effects of an Earthquake, p. 248 ❑ **Interactive Reader and Study Guide**	
❑ **Interactive Reader and Study Guide** ❑ **Directed Reading A** ■ ❑ **Directed Reading B**	❑ TE **Universal Access** Activating Prior Knowledge, p. 249

7

Pacing

- Chapter Lab, Review, and Assessment should take approximately 2.5 days to complete.

Wrapping Up

	Teach
Standards Course of Study	❑ SE **Inquiry Lab** Earthquake Epicenters, pp. 252–253 ❑ **Datasheet B for Chapter Lab** Earthquake Epicenters ■ ❑ **Standards Review Transparency** ■
Universal Access: Advanced Learners/GATE	❑ **Datasheet C for Chapter Lab** Earthquake Epicenters ■
Universal Access: Basic Learners	❑ TE **Universal Access** Preprinted Graphs, p. 254 ❑ TE **Teaching Strategy** Scientific Careers, p. 261 ❑ **Datasheet A for Chapter Lab** Earthquake Epicenters ■
Universal Access: English Learners	❑ TE **Identifying Word Parts** Group Activity, p. 255
Universal Access: Special Education Students	
Universal Access: Struggling Readers	❑ TE **Universal Access** Converting Tables to Graphs, p. 244

Additional Resources

SUPER SUMMARY

Have students review the major concepts in this chapter by using the Super Summary that includes the following:

- an outline of important points in the chapter
- flashcards for chapter vocabulary
- an interactive quiz

Go to **go.hrw.com**
Type in the keyword HY7EQKS

Performance-Based Assessments

The Chapter Resource File for this chapter contains a hands-on activity that can be used to help assess student progress in a nontraditional format. In the Performance-Based Assessment for this chapter, students design and build a structure that will absorb shock from an earthquake.

Key

SE Student Edition
TE Teacher's Edition
Chapter Resource File
Workbook
CD or CD-ROM
Transparency
Video
■ Also available in Spanish

All resources listed below are also available on the One-Stop Planner.

Focus on Earth Sciences: 6.1.a, 6.1.d, 6.1.e, 6.1.g, 6.2.d, 6.3.a, 6.7.c
Math: Algebra and Functions 6.2.3; Mathematical Reasoning 6.2.4
English–Language Arts: Writing 6.1.3

Practice	Assess
❑ SE **Science Skills Activity** Constructing Graphs from Data, p. 254 ❑ **Datasheet for Science Skills Activity** ■ ❑ **Concept Mapping Transparency** ❑ SE **Chapter Review,** pp. 256–257 ❑ **Chapter Review** ■	❑ **Chapter Test B** ■ ❑ SE **Standards Assessment,** pp. 258–259 ❑ **Standards Assessment** ❑ **Standards Review Workbook** ■
	❑ **Chapter Test C** ❑ **Brain Food Video Quiz**
❑ TE **Activity** Researching Local Earthquakes, p. 260	❑ SE **Social Studies Activity,** p. 260 ❑ SE **Math Activity,** p. 261 ❑ **Chapter Test A** ❑ **Brain Food Video Quiz**
❑ TE **Focus on Speaking** Explaining the Standards, p. 255	❑ SE **Language Arts Activity,** p. 260 ❑ **Brain Food Video Quiz**
	❑ **Brain Food Video Quiz**
	❑ **Brain Food Video Quiz**

Holt Online Assessment

Post tests and quizzes to Holt Online Assessment, an assessment management tool. The system automatically grades the assessments, and you receive students' scores and information about which questions students missed. Holt Online Assessment is available through the Premier Online Edition of *Holt California Earth Science.*

Holt Anthology of Science Fiction

The Holt Anthology of Science Fiction includes thought-provoking stories that are relevant to science instruction. Enhance students' learning by asking them to read a story from the *Holt Anthology of Science Fiction* and to answer questions about what they have read.

8 Pacing

- This chapter should take approximately 8 days to complete.
- Getting Started should take approximately 1 day to complete.

Chapter 8 Volcanoes

The Big Idea Volcanoes form as a result of tectonic plate motions and occur where magma reaches Earth's surface.

This chapter was designed to cover the California Grade 6 Science Standards about volcanoes (6.1.a, 6.1.b, 6.1.d, 6.1.e, 6.1.f, 6.2.d, and 6.6.a). This chapter describes volcanoes, their relationship to plate tectonics, and their effects.

After they have completed this chapter, students will begin a unit about processes that shape Earth's surface.

Getting Started

		Teach
	Standards Course of Study	❑ SE **Explore Activity** Predicting a Volcanic Eruption, p. 265 ❑ **Datasheet B for Explore Activity** Predicting a Volcanic Eruption ■
Universal Access	Advanced Learners/GATE	❑ **Chapter Starter Transparency** ❑ **Datasheet C for Explore Activity** Predicting a Volcanic Eruption ■
Universal Access	Basic Learners	❑ SE **Improving Comprehension,** p. 262 ❑ **Chapter Starter Transparency** ❑ **Datasheet A for Explore Activity** Predicting a Volcanic Eruption ■
Universal Access	English Learners	❑ SE **Improving Comprehension,** p. 262 ❑ SE **Unpacking the Standards,** p. 263
Universal Access	Special Education Students	❑ SE **Improving Comprehension,** p. 262 ❑ SE **Unpacking the Standards,** p. 263
Universal Access	Struggling Readers	❑ SE **Improving Comprehension,** p. 262

Key

SE Student Edition | TE Teacher's Edition | Chapter Resource File | Workbook | CD or CD-ROM | Transparency | Video | ■ Also available in Spanish

All resources listed below are also available on the One-Stop Planner.

The California Science Standards listed below are covered in this chapter:

Focus on Earth Sciences

6.1.a Students know evidence of plate tectonics is derived from the fit of the continents; the location of earthquakes, volcanoes, and midocean ridges; and the distribution of fossils, rock types, and ancient climatic zones.

6.1.d Students know that earthquakes are sudden motions along breaks in the crust called faults and that volcanoes and fissures are locations where magma reaches the surface.

6.1.e Students know major geologic events, such as earthquakes, volcanic eruptions, and mountain building, result from plate motions.

6.2.d Students know earthquakes, volcanic eruptions, landslides, and floods change human and wildlife habitats.

6.6.a Students know the utility of energy sources is determined by factors that are involved in converting these sources to useful forms and the consequences of the conversion process.

Investigation and Experimentation

6.7.e Recognize whether evidence is consistent with a proposed explanation.

6.7.h Identify changes in natural phenomena over time without manipulating the phenomena (e.g., a tree limb, a grove of trees, a stream, a hillslope).

Practice	Assess
❑ SE **Organize Activity** Layered Book, p. 264	❑ **Chapter Pretest**
❑ TE **Academic Vocabulary,** p. 263	
❑ TE **Using Other Graphic Organizers,** p. 263	

8 Pacing • This section should take approximately 1 day to complete.

Section 1 Why Volcanoes Form

Key Concept Volcanoes occur at tectonic plate boundaries and at hot spots, where molten rock, or magma, forms and rises to the surface.

	Teach
Standards Course of Study	❑ **Bellringer Transparency** ❑ **PowerPoint® Resources** ❑ SE **Quick Lab** Modeling the Role of Water in Volcanic Eruptions, p. 268 ❑ **Datasheet B for Quick Lab** Modeling the Role of Water in Volcanic Eruptions ■ ❑ **E36** Locations of Earth's Active Volcanoes ❑ **E37** Locations Where Volcanoes Form
Universal Access	
Advanced Learners/GATE	❑ **P14 Link to Physical Science** Changes of State ❑ **Datasheet C for Quick Lab** Modeling the Role of Water in Volcanic Eruptions ■
Basic Learners	❑ TE **Universal Access** Volcanoes of the World, p. 266 ❑ TE **Using the Figure** Volcanoes and Plate Boundaries, p. 266 ❑ **Datasheet A for Quick Lab** Modeling the Role of Water in Volcanic Eruptions ■
English Learners	
Special Education Students	❑ TE **Universal Access** Handy Vocabulary, p. 266
Struggling Readers	❑ SE **Reading Strategy** Summarizing, p. 266

Additional Resources

Holt Lab Generator CD-ROM

Search for any lab by topic, standard, difficulty level, or time. Edit any lab to fit your needs, or create your own labs. Use the Lab Materials QuickList software to customize your lab materials list. Lab datasheets are also available in Spanish on this CD-ROM.

Guided Reading Audio CD Program

The Guided Reading Audio CD Program provides a direct reading of the student text. This resource is helpful to auditory learners and struggling readers. This program is available in English and Spanish.

Key

SE Student Edition | Chapter Resource File | CD or CD-ROM | Video
TE Teacher's Edition | Workbook | Transparency | ■ Also available in Spanish

All resources listed below are also available on the One-Stop Planner.

Focus on Earth Sciences: 6.1.a, 6.1.d, 6.1.e
Math: Algebra and Functions 6.1.2
English–Language Arts: Reading 6.2.4

Practice	Assess
❑ SE **Section Review,** p. 269 ❑ **Section Review** ■	❑ SE **Standards Checks,** pp. 266, 267, 269 ❑ TE **Standards Focus,** p. 268 • Assess • Reteach • Re-Assess ❑ **Section Quiz** ■
❑ TE **Connection to Math** Kilauea, p. 267 ❑ **Interactive Reader and Study Guide** ❑ **Vocabulary and Section Summary A** ■ ❑ **SciLinks Activity** ❑ **Labs You Can Eat** Hot Spots	
❑ **Interactive Reader and Study Guide** ❑ **Vocabulary and Section Summary A** ■ ❑ **Vocabulary and Section Summary B**	❑ TE **Universal Access** Volcano Quizzes, p. 267
❑ **Interactive Reader and Study Guide**	
❑ TE **Universal Access** Retelling Ideas, p. 267 ❑ **Interactive Reader and Study Guide** ❑ **Directed Reading A** ■ ❑ **Directed Reading B**	

Reviewing Prior Knowledge

Prepare students to learn how volcanism is related to tectonic plate motion by reviewing the different types of tectonic plates. See page 266 of the Teacher's Edition.

Nonexplosive Volcanoes Students may think that all volcanic eruptions are explosive and destructive. To correct this misconception, see page 264 in the Teacher's Edition.

Math Support

Science and math go hand in hand. The Math Skills item in the Section Review on page 269 helps students practice math skills in a scientific context.

Pacing • This section should take approximately 2 days to complete.

Section 2 Types of Volcanoes

Key Concept Tectonic plate motions can result in volcanic activity at plate boundaries.

		Teach
	Standards Course of Study	❑ **Bellringer Transparency** ❑ **PowerPoint® Resources** ❑ SE **Quick Lab** Modeling an Explosive Eruption, p. 275 ❑ **Datasheet B for Quick Lab** Modeling an Explosive Eruption ■ ❑ **E38** Four Types of Lava ❑ **E39** Shield Volcano; Cinder Cone; Composite Volcano ❑ **E40** Parts of a Volcano ❑ **E41** Types of Pyroclastic Material
Universal Access	**Advanced Learners/GATE**	❑ **Datasheet C for Quick Lab** Modeling an Explosive Eruption ■
Universal Access	**Basic Learners**	❑ TE **Discussion** Student Impressions of Eruptions, p. 270 ❑ TE **Group Activity** Describing Viscosity, p. 273 ❑ **Datasheet A for Quick Lab** Modeling an Explosive Eruption ■
Universal Access	**English Learners**	❑ TE **Wordwise** Word Connections, p. 275
Universal Access	**Special Education Students**	❑ TE **Universal Access** Going Global, p. 271
Universal Access	**Struggling Readers**	❑ SE **Reading Strategy** Graphic Organizer, p. 270

Key

SE Student Edition
TE Teacher's Edition

Chapter Resource File
Workbook

CD or CD-ROM
Transparency

Video
■ Also available in Spanish

All resources listed below are also available on the One-Stop Planner.

Focus on Earth Sciences: 6.1.d, 6.1.e
Math: Algebra and Functions 6.2.1; Number Sense 6.1.3

Practice	Assess
❑ SE **Section Review,** p. 277 ❑ **Section Review** ■	❑ SE **Standards Checks,** pp. 270, 271, 272, 273, 274, 276 ❑ TE **Standards Focus,** p. 276 • Assess • Reteach • Re-Assess ❑ **Section Quiz** ■
❑ TE **Activity** Book Report, p. 273	❑ **Long-Term Projects & Research Ideas** Legend Has It ...
❑ TE **Group Activity** Making Lava, p. 274 ❑ TE **Universal Access** Volcano Field Guide, p. 275 ❑ **Interactive Reader and Study Guide** ❑ **Vocabulary and Section Summary A** ■ ❑ **Reinforcement Worksheet**	
❑ TE **Universal Access** Volcanoes in Depth, p. 274 ❑ **Interactive Reader and Study Guide** ❑ **Vocabulary and Section Summary A** ■ ❑ **Vocabulary and Section Summary B**	❑ TE **Activity** Classifying Volcanoes, p. 271
❑ TE **Universal Access** Lava Models, p. 272 ❑ **Interactive Reader and Study Guide**	
❑ TE **Universal Access** Scanning for New Vocabulary, p. 270 ❑ **Interactive Reader and Study Guide** ❑ **Directed Reading A** ■ ❑ **Directed Reading B**	

8

Pacing

- This section should take approximately 1 day to complete.

Section 3 Effects of Volcanic Eruptions

Key Concept The effects of volcanic eruptions can change human and wildlife habitats.

	Teach
Standards Course of Study	❑ **Bellringer Transparency** ❑ **PowerPoint® Resources** ❑ SE **Quick Lab** Modeling Ash and Gases in Earth's Atmosphere, p. 279 ❑ **Datasheet B for Quick Lab** Modeling Ash and Gases in Earth's Atmosphere ■
Universal Access: Advanced Learners/GATE	❑ **Datasheet C for Quick Lab** Modeling Ash and Gases in Earth's Atmosphere ■
Universal Access: Basic Learners	❑ **Datasheet A for Quick Lab** Modeling Ash and Gases in Earth's Atmosphere ■
Universal Access: English Learners	
Universal Access: Special Education Students	
Universal Access: Struggling Readers	❑ SE **Reading Strategy** Graphic Organizer, p. 278 ❑ TE **Universal Access** Predicting, p. 279

Additional Resources

Holt Lab Generator CD-ROM

Search for any lab by topic, standard, difficulty level, or time. Edit any lab to fit your needs, or create your own labs. Use the Lab Materials QuickList software to customize your lab materials list. Lab datasheets are also available in Spanish on this CD-ROM.

Guided Reading Audio CD Program

The Guided Reading Audio CD Program provides a direct reading of the student text. This resource is helpful to auditory learners and struggling readers. This program is available in English and Spanish.

Key

SE Student Edition
TE Teacher's Edition
Chapter Resource File
Workbook
CD or CD-ROM
Transparency
Video
■ Also available in Spanish

All resources listed below are also available on the One-Stop Planner.

Focus on Earth Sciences: 6.2.d, 6.6.a
Math: Algebra and Functions 6.2.2

Practice	Assess
❑ SE **Section Review,** p. 281 ❑ **Section Review** ■	❑ SE **Standards Checks,** pp. 279, 280 ❑ TE **Standards Focus,** p. 280 • Assess • Reteach • Re-Assess ❑ **Section Quiz** ■
❑ TE **Universal Access** Mapping Volcanoes, p. 278 ❑ **Critical Thinking**	❑ **Long-Term Projects & Research Ideas** A City Lost and Found
❑ TE **Group Activity** Preparing for an Eruption, p. 278 ❑ **Interactive Reader and Study Guide** ❑ **Vocabulary and Section Summary A** ■	❑ TE **Universal Access** Volcano Pen Pals, p. 278
❑ TE **Universal Access** Illustrating Effects, p. 280 ❑ **Interactive Reader and Study Guide** ❑ **Vocabulary and Section Summary A** ■ ❑ **Vocabulary and Section Summary B**	❑ TE **Universal Access** Eruption Pros and Cons, p. 280
❑ **Interactive Reader and Study Guide**	
❑ **Interactive Reader and Study Guide** ❑ **Directed Reading A** ■ ❑ **Directed Reading B**	

Reviewing Prior Knowledge

Prepare students to learn the ways in which volcanic eruptions change human and wildlife habitats. See page 278 of the Teacher's Edition.

Nonexplosive Eruptions Students may think that explosive volcanic eruptions play a greater role in shaping our world than nonexplosive eruptions do. To correct this misconception, see page 279 in the Teacher's Edition.

8 **Pacing**

- Chapter Lab, Review, and Assessment should take approximately 4 days to complete.

Wrapping Up

	Teach
Standards Course of Study	❑ SE **Skills Practice Lab** Locating Earth's Volcanoes, pp. 282–283 ❑ **Datasheet B for Chapter Lab** Locating Earth's Volcanoes ■ ❑ **Standards Review Transparency** ■
Universal Access: Advanced Learners/GATE	❑ **Datasheet C for Chapter Lab** Locating Earth's Volcanoes ■
Universal Access: Basic Learners	❑ TE **Identifying Suffixes** Group Activity, p. 285 ❑ **Datasheet A for Chapter Lab** Locating Earth's Volcanoes ■
Universal Access: English Learners	
Universal Access: Special Education Students	
Universal Access: Struggling Readers	

Additional Resources

SUPER SUMMARY

Have students review the major concepts in this chapter by using the Super Summary that includes the following:

- an outline of important points in the chapter
- flashcards for chapter vocabulary
- an interactive quiz

Go to **go.hrw.com**
Type in the keyword HY7VOLS

Performance-Based Assessments

The Chapter Resource File for this chapter contains a hands-on activity that can be used to help assess student progress in a nontraditional format. In the Performance-Based Assessment for this chapter, students model lava flows.

Key

SE Student Edition
TE Teacher's Edition
Chapter Resource File
Workbook
CD or CD-ROM
Transparency
Video
■ Also available in Spanish

All resources listed below are also available on the One-Stop Planner.

Focus on Earth Sciences: 6.1.a, 6.1.d, 6.1.e, 6.2.d, 6.6.a, 6.7.h
Math: Algebra and Functions 6.2.2
English–Language Arts: Writing 6.1.2

Practice	Assess
❑ SE **Science Skills Activity** Constructing Graphs from Data, p. 284 ❑ **Datasheet for Science Skills Activity** ■ ❑ **Concept Mapping Transparency** ❑ SE **Chapter Review,** pp. 286–287 ❑ **Chapter Review** ■	❑ **Chapter Test B** ■ ❑ SE **Standards Assessment,** pp. 288–289 ❑ **Standards Assessment** ❑ **Standards Review Workbook** ■
	❑ **Chapter Test C** ❑ **Brain Food Video Quiz**
	❑ SE **Social Studies Activity,** p. 290 ❑ SE **Math Activity,** p. 291 ❑ **Chapter Test A** ❑ **Brain Food Video Quiz**
❑ TE **Focus on Writing** Connecting Concepts, p. 285	❑ SE **Language Arts Activity,** p. 290 ❑ **Brain Food Video Quiz**
❑ TE **Universal Access** Support Questions, p. 284	❑ **Brain Food Video Quiz**
❑ TE **Universal Access** Comparing and Contrasting, p. 284	❑ **Brain Food Video Quiz**

Holt Online Assessment

Post tests and quizzes to Holt Online Assessment, an assessment management tool. The system automatically grades the assessments, and you receive students' scores and information about which questions students missed. Holt Online Assessment is available through the Premier Online Edition of *Holt California Earth Science.*

Holt Anthology of Science Fiction

The Holt Anthology of Science Fiction includes thought-provoking stories that are relevant to science instruction. Enhance students' learning by asking them to read a story from the *Holt Anthology of Science Fiction* and to answer questions about what they have read.

9

Pacing

- This chapter should take approximately 10.5 days to complete.
- Getting Started should take approximately 1 day to complete.

Chapter 9 Weathering and Soil Formation

The Big Idea Weathering is a continuous process that results in the formation of soil and the construction and destruction of landforms.

This chapter was designed to cover the California Grade 6 Science Standards about weathering of rocks and about soil as a resource (6.2.a, 6.2.b, 6.2.c, 6.5.e, and 6.6.b). The chapter discusses the historical development of soil science and the discussion of environmental science and the conservation of resources as referred to in Category 1, Criterion 10 and in Category 1, Criterion 11 of the Criteria for Evaluating Instructional Materials in Science. This chapter is part of a unit that covers ways in which Earth's surface is shaped. This chapter describes the continuous weathering process and its role in soil formation and shaping landforms.

After they have completed this chapter, students will begin a chapter about the various agents of erosion and deposition.

Getting Started

		Teach
	Standards Course of Study	❑ **SE Explore Activity** Break It Down, p. 297 ❑ **Datasheet B for Explore Activity** Break It Down ■
Universal Access	Advanced Learners/GATE	❑ **Chapter Starter Transparency** ❑ **Datasheet C for Explore Activity** Break It Down ■
	Basic Learners	❑ **SE Improving Comprehension,** p. 294 ❑ **Chapter Starter Transparency** ❑ **Datasheet A for Explore Activity** Break It Down ■
	English Learners	❑ **SE Improving Comprehension,** p. 294 ❑ **SE Unpacking the Standards,** p. 295
	Special Education Students	❑ **SE Improving Comprehension,** p. 294 ❑ **SE Unpacking the Standards,** p. 295
	Struggling Readers	❑ **SE Improving Comprehension,** p. 294

Key

SE Student Edition	Chapter Resource File	CD or CD-ROM
TE Teacher's Edition	Workbook	Transparency
Video	■ Also available in Spanish	

All resources listed below are also available on the One-Stop Planner.

The California Science Standards listed below are covered in this chapter:

Focus on Earth Sciences

6.2.a Students know water running downhill is the dominant process in shaping the landscape, including California's landscape.

6.2.b Students know rivers and streams are dynamic systems that erode, transport sediment, change course, and flood their banks in natural and recurring patterns.

6.2.c Students know beaches are dynamic systems in which the sand is supplied by rivers and moved along the coast by the action of waves.

6.5.e Students know the number and types of organisms an ecosystem can support depends on the resources available and on abiotic factors, such as quantities of light and water, a range of temperatures, and soil composition.

6.6.b Students know different natural energy and material resources, including air, soil, rocks, minerals, petroleum, fresh water, wildlife, and forests, and know how to classify them as renewable or nonrenewable.

Investigation and Experimentation

6.7.a Develop a hypothesis.

6.7.c Construct appropriate graphs from data and develop qualitative statements about the relationships between variables.

Practice	Assess
❑ **SE Organize Activity** Key-Term Fold, p. 296	❑ **Chapter Pretest**
❑ **TE Words with Multiple Meanings,** p. 295	
❑ **TE Using Other Graphic Organizers,** p. 294	

9 Pacing • This section should take approximately 1.5 days to complete.

Section 1 Weathering

Key Concept Rock is broken down into smaller pieces by mechanical and chemical weathering.

	Teach
Standards Course of Study	❑ **Bellringer Transparency** ❑ **PowerPoint® Resources** ❑ **E42** Chemical Weathering of Granite ❑ **SE Quick Lab** The Reaction of Acids, p. 302 ❑ **Datasheet B for Quick Lab** The Reaction of Acids ■
Universal Access: Advanced Learners/GATE	❑ **Datasheet C for Quick Lab** The Reaction of Acids ■
Universal Access: Basic Learners	❑ **Datasheet A for Quick Lab** The Reaction of Acids ■
Universal Access: English Learners	❑ **TE Demonstration** CO_2 and Rain, p. 301
Universal Access: Special Education Students	❑ **TE Universal Access** Rock Sorting, p. 298
Universal Access: Struggling Readers	❑ **SE Reading Strategy** Graphic Organizer, p. 298 ❑ **TE Universal Access** Previewing and Predicting, p. 298

Additional Resources

Holt Lab Generator CD-ROM

Search for any lab by topic, standard, difficulty level, or time. Edit any lab to fit your needs, or create your own labs. Use the Lab Materials QuickList software to customize your lab materials list. Lab datasheets are also available in Spanish on this CD-ROM.

Guided Reading Audio CD Program

The Guided Reading Audio CD Program provides a direct reading of the student text. This resource is helpful to auditory learners and struggling readers. This program is available in English and Spanish.

Key

SE Student Edition
TE Teacher's Edition
Chapter Resource File
Workbook
CD or CD-ROM
Transparency
Video
■ Also available in Spanish

All resources listed below are also available on the One-Stop Planner.

Focus on Earth Sciences: 6.2.a, 6.2.b
Math: Number Sense 6.2.1

Practice	Assess
❑ SE **Section Review,** p. 303 ❑ **Section Review** ■	❑ SE **Standards Checks,** pp. 299, 300, 303 ❑ TE **Standards Focus,** p. 302 • Assess • Reteach • Re-Assess ❑ **Section Quiz** ■
	❑ TE **Universal Access** Weathering Poster, p. 300
❑ TE **Group Activity** Identifying Weathering, p. 298 ❑ **Interactive Reader and Study Guide** ❑ **Vocabulary and Section Summary A** ■ ❑ **Reinforcement Worksheet**	
❑ **Interactive Reader and Study Guide** ❑ **Vocabulary and Section Summary A** ■ ❑ **Vocabulary and Section Summary B**	❑ TE **Universal Access** Demonstrating Weathering, p. 301 ❑ TE **Group Activity** Acid Precipitation, p. 301
❑ **Interactive Reader and Study Guide**	
❑ **Interactive Reader and Study Guide** ❑ **Directed Reading A** ■ ❑ **Directed Reading B**	

Reviewing Prior Knowledge

Prepare students to learn about the effects of weathering of rocks by reviewing topographic maps. See p. 298 of the Teacher's Edition.

Humans Cause Weathering Students may think that humans do not contribute to weathering. To correct this misconception, see p. 300 in the Teacher's Edition.

Math Support

Science and math go hand in hand. The Math Skills item in the Section Review on p. 303 helps students practice math skills in a scientific context.

9 **Pacing** • **This section should take approximately 1.5 days to complete.**

Section 2 Rates of Weathering

Key Concept The rate at which rock weathers depends on climate, elevation, and the size and makeup of the rock.

	Teach
Standards Course of Study	❑ **Bellringer Transparency** ❑ **PowerPoint® Resources** ❑ SE **Quick Lab** How Fast Will It Dissolve?, p. 306 ❑ **Datasheet B for Quick Lab** How Fast Will It Dissolve? ■
Universal Access — Advanced Learners/GATE	❑ **Datasheet C for Quick Lab** How Fast Will It Dissolve? ■ ❑ **L10 Link to Life Science** Math Focus: Surface Area-to-Volume Ratio
Universal Access — Basic Learners	❑ TE **Universal Access** Differential Weathering, p. 304 ❑ **Datasheet A for Quick Lab** How Fast Will It Dissolve? ■
Universal Access — English Learners	❑ TE **Group Activity** Surface Area and Weathering, p. 304 ❑ TE **Using the Figure** Surface Area and Volume, p. 305 ❑ TE **Universal Access** Understanding Word Families, p. 304
Universal Access — Special Education Students	❑ TE **Universal Access** Weathering Potatoes, p. 305
Universal Access — Struggling Readers	❑ SE **Reading Strategy** Outlining, p. 304

Additional Resources

Holt Lab Generator CD-ROM

Search for any lab by topic, standard, difficulty level, or time. Edit any lab to fit your needs, or create your own labs. Use the Lab Materials QuickList software to customize your lab materials list. Lab datasheets are also available in Spanish on this CD-ROM.

Guided Reading Audio CD Program

The Guided Reading Audio CD Program provides a direct reading of the student text. This resource is helpful to auditory learners and struggling readers. This program is available in English and Spanish.

Key

SE Student Edition
TE Teacher's Edition
Chapter Resource File
Workbook
CD or CD-ROM
Transparency
Video
■ Also available in Spanish

All resources listed below are also available on the One-Stop Planner.

Focus on Earth Sciences: 6.2.c
English–Language Arts: Reading 6.2.4

Practice	Assess
❑ SE **Section Review,** p. 307 ❑ **Section Review** ■	❑ SE **Standards Check,** p. 304 ❑ TE **Standards Focus,** p. 306 • Assess • Reteach • Re-Assess ❑ **Section Quiz** ■
❑ **Interactive Reader and Study Guide** ❑ **Vocabulary and Section Summary A** ■	
❑ **Interactive Reader and Study Guide** ❑ **Vocabulary and Section Summary A** ■ ❑ **Vocabulary and Section Summary B**	
❑ **Interactive Reader and Study Guide**	
❑ **Interactive Reader and Study Guide** ❑ **Directed Reading A** ■ ❑ **Directed Reading B**	

Reviewing Prior Knowledge

Prepare students to learn about rates of weathering by reviewing what they know about the rock cycle. See p. 304 of the Teacher's Edition.

MISCONCEPTION ALERT

Weathering of Hard Rocks Students may think that some rocks do not weather. To correct this misconception, see p. 305 in the Teacher's Edition.

9 Pacing • This section should take approximately 1.5 days to complete.

Section 3 From Bedrock to Soil

Key Concept Weathering may lead to the formation of soil, which is an important natural resource.

		Teach
	Standards Course of Study	❑ **Bellringer Transparency** ❑ **PowerPoint® Resources** ❑ **E43** Soil Horizons ❑ **E44** pH Values of Common Materials ❑ **SE Quick Lab** Investigating Plant Growth, p. 311 ❑ **Datasheet B for Quick Lab** Investigating Plant Growth ■
Universal Access	Advanced Learners/GATE	❑ **Datasheet C for Quick Lab** Investigating Plant Growth ■
Universal Access	Basic Learners	❑ **Datasheet A for Quick Lab** Investigating Plant Growth ■
Universal Access	English Learners	
Universal Access	Special Education Students	
Universal Access	Struggling Readers	❑ **SE Reading Strategy** Summarizing, p. 308 ❑ **TE Reading Strategy** Anticipation Guide, p. 309 ❑ **TE Universal Access** Figure References, p. 312

Additional Resources

Holt Lab Generator CD-ROM

Search for any lab by topic, standard, difficulty level, or time. Edit any lab to fit your needs, or create your own labs. Use the Lab Materials QuickList software to customize your lab materials list. Lab datasheets are also available in Spanish on this CD-ROM.

Guided Reading Audio CD Program

The Guided Reading Audio CD Program provides a direct reading of the student text. This resource is helpful to auditory learners and struggling readers. This program is available in English and Spanish.

Key

SE Student Edition		**CD or CD-ROM**	 **Video**
TE Teacher's Edition	**Workbook**	**Transparency**	■ **Also available in Spanish**

Chapter Resource File

All resources listed below are also available on the One-Stop Planner.

Focus on Earth Sciences: 6.5.e
Math: Number Sense 6.2.1
English–Language Arts: Reading 6.2.4

Practice	Assess
❑ **SE Section Review,** p. 313 ❑ **Section Review** ■	❑ **SE Standards Checks,** pp. 309, 311, 313 ❑ **TE Standards Focus,** p. 312 • Assess • Reteach • Re-Assess ❑ **Section Quiz** ■
❑ **Calculator-Based Lab** A Soil Study ❑ **Long-Term Projects & Research Ideas** Precious Soil	❑ **TE Homework** Desert Pavement, p. 309 ❑ **TE Universal Access** Studying Local Soils, p. 310
❑ **TE Universal Access** Outlining Concepts, p. 309 ❑ **TE Connection to Biology** Berlese Funnels, p. 310 ❑ **Interactive Reader and Study Guide** ❑ **Vocabulary and Section Summary A** ■	
❑ **TE Group Activity** Describing Soil, p. 308 ❑ **TE Universal Access** Comparing Properties, p. 308 ❑ **TE Group Activity** Living Soil, p. 310 ❑ **Interactive Reader and Study Guide** ❑ **Vocabulary and Section Summary A** ■ ❑ **Vocabulary and Section Summary B**	
❑ **TE Universal Access** Dirty Work, p. 309 ❑ **Interactive Reader and Study Guide**	
❑ **Interactive Reader and Study Guide** ❑ **Directed Reading A** ■ ❑ **Directed Reading B**	

Reviewing Prior Knowledge

Prepare students to learn about soil by having students predict the relationship between soil and ecosystems. See p. 308 of the Teacher's Edition.

Understanding pH Students may think that a decrease in pH indicates only a slightly higher acidity. To correct this misconception, see p. 311 in the Teacher's Edition.

Math Support

Science and math go hand in hand. The Math Practice on p. 309 helps students practice math skills in a scientific context.

9 **Pacing** • This section should take approximately 1 day to complete.

Section 4 Soil Conservation

Key Concept Soil is a nonrenewable resource that can be endangered if it is used unwisely.

	Teach
Standards Course of Study	❑ **Bellringer Transparency** ❑ **PowerPoint® Resources** ❑ SE **Quick Lab** Soil Erosion, p. 315 ❑ **Datasheet B for Quick Lab** Soil Erosion ■
Universal Access: Advanced Learners/GATE	❑ **Datasheet C for Quick Lab** Soil Erosion ■
Universal Access: Basic Learners	❑ **Datasheet A for Quick Lab** Soil Erosion ■
Universal Access: English Learners	
Universal Access: Special Education Students	
Universal Access: Struggling Readers	❑ SE **Reading Strategy** Graphic Organizer, p. 314 ❑ TE **Universal Access** Activating Prior Knowledge, p. 314

Additional Resources

Holt Lab Generator CD-ROM

Search for any lab by topic, standard, difficulty level, or time. Edit any lab to fit your needs, or create your own labs. Use the Lab Materials QuickList software to customize your lab materials list. Lab datasheets are also available in Spanish on this CD-ROM.

Guided Reading Audio CD Program

The Guided Reading Audio CD Program provides a direct reading of the student text. This resource is helpful to auditory learners and struggling readers. This program is available in English and Spanish.

Key

SE Student Edition
TE Teacher's Edition
Chapter Resource File
Workbook
CD or CD-ROM
Transparency

Video
■ Also available in Spanish

All resources listed below are also available on the One-Stop Planner.

Focus on Earth Sciences: 6.5.e
Math: Number Sense 6.2.1
English–Language Arts: Reading 6.2.4

Practice	Assess
❑ SE **Section Review,** p. 317 ❑ **Section Review** ■	❑ SE **Standards Checks,** pp. 314, 317 ❑ TE **Standards Focus,** p. 318 • Assess • Reteach • Re-Assess ❑ **Section Quiz** ■
❑ TE **Universal Access** Looking for Erosion, p. 314 ❑ **Critical Thinking** ❑ **SciLinks Activity**	
❑ **Interactive Reader and Study Guide** ❑ **Vocabulary and Section Summary A** ■	
❑ **Interactive Reader and Study Guide** ❑ **Vocabulary and Section Summary A** ■ ❑ **Vocabulary and Section Summary B**	❑ TE **Universal Access** Public Service Announcements, p. 314
❑ **Interactive Reader and Study Guide**	
❑ **Interactive Reader and Study Guide** ❑ **Directed Reading A** ■ ❑ **Directed Reading B**	

Reviewing Prior Knowledge

Prepare students to learn about soil conservation by discussing the Dust Bowl era. See p. 314 of the Teacher's Edition.

Pores in Soil Students may not be aware of the importance of pore spaces in soil. To correct this misconception, see p. 315 in the Teacher's Edition.

Math Support

Science and math go hand in hand. The Math Skills item in the Section Review on p. 317 helps students practice math skills in a scientific context.

9 Pacing

- Chapter Lab, Review, and Assessment should take approximately 4 days to complete.

Wrapping Up

	Teach
Standards Course of Study	❑ SE **Model-Making Lab** Weathering Rocks, pp. 318–319 ❑ **Datasheet B for Chapter Lab** Weathering Rocks ■ ❑ **Standards Review Transparency** ■
Universal Access: Advanced Learners/GATE	❑ **Datasheet C for Chapter Lab** Weathering Rocks ■
Universal Access: Basic Learners	❑ **Datasheet A for Chapter Lab** Weathering Rocks ■ ❑ TE **Universal Access** Forming Hypotheses, p. 320 ❑ TE **Teaching Strategy,** p. 327
Universal Access: English Learners	
Universal Access: Special Education Students	❑ TE **Universal Access** Stop and Discuss, p. 320
Universal Access: Struggling Readers	

Additional Resources

SUPER SUMMARY

Have students review the major concepts in this chapter by using the Super Summary that includes the following:

- an outline of important points in the chapter
- flashcards for chapter vocabulary
- an interactive quiz

Go to **go.hrw.com**
Type in the keyword HY7WSFS

Performance-Based Assessments

The Chapter Resource File for this chapter contains a hands-on activity that can be used to help assess student progress in a nontraditional format. In the Performance-Based Assessment for this chapter, students model differential weathering.

Key

SE Student Edition
TE Teacher's Edition
Chapter Resource File
Workbook
CD or CD-ROM
Transparency
Video
■ Also available in Spanish

All resources listed below are also available on the One-Stop Planner.

Focus on Earth Sciences: 6.2.a, 6.2.b, 6.2.c, 6.5.e, 6.6.b, 6.7.a, 6.7.c
Math: Number Sense 6.2.1
English–Language Arts: Writing 6.1.3

Practice	Assess
❑ **Concept Mapping Transparency** ❑ SE **Science Skills Activity** Constructing Graphs from Data, p. 320 ❑ SE **Chapter Review,** pp. 322–323 ❑ **Datasheet for Science Skills Activity** ■ ❑ **Chapter Review** ■	❑ **Chapter Test B** ■ ❑ SE **Standards Assessment,** pp. 324–325 ❑ **Standards Assessment** ❑ **Standards Review Workbook** ■
	❑ SE **Language Arts Activity,** p. 326 ❑ TE **Focus on Speaking,** p. 321 ❑ **Chapter Test C** ❑ **Brain Food Video Quiz**
❑ TE **Identifying Roots,** p. 321 ❑ TE **Activity,** p. 326	❑ SE **Social Studies Activity,** p. 326 ❑ SE **Math Activity,** p. 327 ❑ **Chapter Test A** ❑ **Brain Food Video Quiz**
	❑ **Brain Food Video Quiz**
	❑ **Brain Food Video Quiz**
	❑ **Brain Food Video Quiz**

Holt Online Assessment

Post tests and quizzes to Holt Online Assessment, an assessment management tool. The system automatically grades the assessments, and you receive students' scores and information about which questions students missed. Holt Online Assessment is available through the Premier Online Edition of *Holt California Earth Science.*

Holt Anthology of Science Fiction

The Holt Anthology of Science Fiction includes thought-provoking stories that are relevant to science instruction. Enhance students' learning by asking them to read a story from the *Holt Anthology of Science Fiction* and to answer questions about what they have read.

10 **Pacing**
- **This chapter should take approximately 10 days to complete.**
- **Getting Started should take approximately 1 day to complete.**

Chapter 10 Agents of Erosion and Deposition

The Big Idea Topography is reshaped by the weathering of rock and soil and by the transportation and deposition of sediment.

This chapter was designed to cover the California Grade 6 Science Standards about erosion and deposition (6.2.a, 6.2.c, 6.2.d, and 6.3.a). It follows a chapter about weathering and soil formation. This chapter describes the various agents of erosion and deposition and their effect on topography.

After they have completed this chapter, students will begin a chapter about rivers and groundwater.

Getting Started

		Teach
	Standards Course of Study	❑ **SE Explore Activity** Shaping Beaches by Wave Erosion, p. 331 ❑ **Datasheet B for Explore Activity** Shaping Beaches by Wave Erosion ■
Universal Access	Advanced Learners/GATE	❑ **Chapter Starter Transparency** ❑ **Datasheet C for Explore Activity** Shaping Beaches by Wave Erosion ■
Universal Access	Basic Learners	❑ **SE Improving Comprehension,** p. 328 ❑ **Chapter Starter Transparency** ❑ **Datasheet A for Explore Activity** Shaping Beaches by Wave Erosion ■
Universal Access	English Learners	❑ **SE Improving Comprehension,** p. 328 ❑ **SE Unpacking the Standards,** p. 329
Universal Access	Special Education Students	❑ **SE Improving Comprehension,** p. 328 ❑ **SE Unpacking the Standards,** p. 329
Universal Access	Struggling Readers	❑ **SE Improving Comprehension,** p. 328

Key

SE Student Edition
TE Teacher's Edition

Chapter Resource File
Workbook

CD or CD-ROM
Transparency

Video
Also available in Spanish

All resources listed below are also available on the One-Stop Planner.

The California Science Standards listed below are covered in this chapter:

Focus on Earth Sciences

6.2.a Students know water running downhill is the dominant process in shaping the landscape, including California's landscape.

6.2.c Students know beaches are dynamic systems in which the sand is supplied by rivers and moved along the coast by the action of waves.

6.2.d Students know earthquakes, volcanic eruptions, landslides, and floods change human and wildlife habitats.

6.3.a Students know energy can be carried from one place to another by heat flow or by waves, including water, light and sound waves, or by moving objects.

Investigation and Experimentation

6.7.a Develop a hypothesis.

6.7.b Select and use appropriate tools and technology (including calculators, computers, balances, spring scales, microscopes, and binoculars) to perform tests, collect data, and display data.

6.7.e Recognize whether evidence is consistent with a proposed explanation.

Practice	Assess
❑ SE **Organize Activity** Layered Book, p. 330	❑ **Chapter Pretest**
❑ TE **Words with Multiple Meanings,** p. 329	
❑ TE **Using Other Graphic Organizers,** p. 328	

10 **Pacing** • This section should take approximately 2 days to complete.

Section 1 Shoreline Erosion and Deposition

Key Concept Beaches and shorelines are shaped largely by the action of ocean waves.

	Teach
Standards Course of Study	❑ **Bellringer Transparency** ❑ **PowerPoint® Resources** ❑ **E45** Coastal Landforms Created by Wave Erosion: A ❑ **E46** Coastal Landforms Created by Wave Erosion: B ❑ **E47** Longshore Current ❑ SE **Quick Lab** Observing Differences in Sand, p. 338 ❑ **Datasheet B for Quick Lab** Observing Differences in Sand ■
Universal Access: Advanced Learners/GATE	❑ **P46 Link to Physical Science** Static Friction ❑ **Datasheet C for Quick Lab** Observing Differences in Sand ■
Universal Access: Basic Learners	❑ TE **Discussion** Insuring Ocean Properties, p. 335 ❑ **Datasheet A for Quick Lab** Observing Differences in Sand ■
Universal Access: English Learners	❑ TE **Demonstration** Wave Deposits, p. 337
Universal Access: Special Education Students	❑ TE **Universal Access** An Oral Visual, p. 334
Universal Access: Struggling Readers	❑ SE **Reading Strategy** Prediction Guide, p. 332 ❑ TE **Universal Access** Understanding Roots, p. 333 ❑ TE **Universal Access** Activating Prior Knowledge, p. 336

Additional Resources

Holt Lab Generator CD-ROM

Search for any lab by topic, standard, difficulty level, or time. Edit any lab to fit your needs, or create your own labs. Use the Lab Materials QuickList software to customize your lab materials list. Lab datasheets are also available in Spanish on this CD-ROM.

Guided Reading Audio CD Program

The Guided Reading Audio CD Program provides a direct reading of the student text. This resource is helpful to auditory learners and struggling readers. This program is available in English and Spanish.

Key

SE Student Edition
TE Teacher's Edition
Chapter Resource File
Workbook
CD or CD-ROM
Transparency
Video
■ Also available in Spanish

All resources listed below are also available on the One-Stop Planner.

Focus on Earth Sciences: 6.2.c, 6.3.a
Math: Number Sense 6.2.1

Practice	Assess
❑ SE **Section Review,** p. 339 ❑ **Section Review** ■	❑ SE **Standards Checks,** pp. 333, 334, 335, 336, 337 ❑ TE **Standards Focus,** p. 338 • Assess • Reteach • Re-Assess ❑ **Section Quiz** ■
❑ **Interactive Reader and Study Guide** ❑ **Vocabulary and Section Summary A** ■ ❑ **Inquiry Labs** Surf's Up!	❑ TE **Activity** Illustrating Beach Erosion, p. 332
❑ TE **Group Activity** Coastal Features Board Game, p. 334 ❑ TE **Group Activity** Waves, p. 336 ❑ TE **Using the Figure** Beach Materials, p. 336 ❑ **Interactive Reader and Study Guide** ❑ **Vocabulary and Section Summary A** ■ ❑ **Vocabulary and Section Summary B**	❑ TE **Universal Access** Reviewing Key Vocabulary, p. 337
❑ **Interactive Reader and Study Guide**	❑ TE **Universal Access** Beach Posters, p. 336
❑ **Interactive Reader and Study Guide** ❑ **Directed Reading A** ■ ❑ **Directed Reading B**	

Reviewing Prior Knowledge

Prepare students to learn about how shorelines are shaped by reviewing how waves transfer energy from one place to another. See p. 332 of the Teacher's Edition.

Wave Movement Students may think that a wave is a moving wall of water. To correct this misconception, see p. 333 in the Teacher's Edition.

Math Support

Science and math go hand in hand. The Math Skills item in the Section Review on p. 339 helps students practice math skills in a scientific context.

10 Pacing • This section should take approximately 1 day to complete.

Section 2 Wind Erosion and Deposition

Key Concept Wind can cause erosion and can move and deposit sediment.

	Teach
Standards Course of Study	❑ **Bellringer Transparency** ❑ **PowerPoint® Resources** ❑ **E48** The Movement of Dunes ❑ **SE Quick Lab** Making Desert Pavement, p. 341 ❑ **Datasheet B for Quick Lab** Making Desert Pavement ■
Universal Access: Advanced Learners/GATE	❑ **Datasheet C for Quick Lab** Making Desert Pavement ■
Universal Access: Basic Learners	❑ TE **Discussion Wind Erosion,** p. 340 ❑ **Datasheet A for Quick Lab** Making Desert Pavement ■
Universal Access: English Learners	❑ TE **Universal Access** Demonstrating Deflation, p. 341
Universal Access: Special Education Students	
Universal Access: Struggling Readers	❑ **SE Reading Strategy** Graphic Organizer, p. 340 ❑ TE **Universal Access** Understanding Vocabulary, p. 340

Additional Resources

Holt Lab Generator CD-ROM

Search for any lab by topic, standard, difficulty level, or time. Edit any lab to fit your needs, or create your own labs. Use the Lab Materials QuickList software to customize your lab materials list. Lab datasheets are also available in Spanish on this CD-ROM.

Guided Reading Audio CD Program

The Guided Reading Audio CD Program provides a direct reading of the student text. This resource is helpful to auditory learners and struggling readers. This program is available in English and Spanish.

Key
SE Student Edition
TE Teacher's Edition

Chapter Resource File
Workbook
CD or CD-ROM
Transparency

Video
■ Also available in Spanish

All resources listed below are also available on the One-Stop Planner.

Focus on Earth Sciences: 6.2.a, 6.7.e
English–Language Arts: Reading 6.2.4

Practice	Assess
❑ SE **Section Review,** p. 343 ❑ **Section Review** ■	❑ SE **Standards Checks,** pp. 341, 343 ❑ TE **Standards Focus,** p. 342 • Assess • Reteach • Re-Assess ❑ **Section Quiz** ■
❑ **Interactive Reader and Study Guide** ❑ **Vocabulary and Section Summary A** ■	❑ TE **Universal Access** Life in the Desert, p. 342
❑ TE **Wordwise Word Connections,** p. 341 ❑ **Interactive Reader and Study Guide** ❑ **Vocabulary and Section Summary A** ■ ❑ **Vocabulary and Section Summary B**	
❑ TE **Universal Access** Making Memories, p. 341 ❑ **Interactive Reader and Study Guide**	
❑ **Interactive Reader and Study Guide** ❑ **Directed Reading A** ■ ❑ **Directed Reading B**	

Reviewing Prior Knowledge

Prepare students to learn about wind erosion and deposition by reviewing the definitions of vocabulary words. See p. 340 of the Teacher's Edition.

Deflation Students may be unclear about the meaning of *deflation.* To correct this misconception, see p. 340 in the Teacher's Edition.

Pacing

- This section should take approximately 1 day to complete.

Section 3 Erosion and Deposition by Ice

Key Concept Glaciers shape Earth's surface by moving rock and sediment.

	Teach
Standards Course of Study	❑ **Bellringer Transparency** ❑ **PowerPoint® Resources** ❑ **E49** Landscape Features Carved by Alpine Glaciers ❑ SE **Quick Lab** Modeling a Glacier, p. 346 ❑ **Datasheet B for Quick Lab** Modeling a Glacier ■
Universal Access	
Advanced Learners/GATE	❑ TE **Universal Access** Evidence of Glaciers, p. 344 ❑ **Datasheet C for Quick Lab** Modeling a Glacier ■
Basic Learners	❑ **Datasheet A for Quick Lab** Modeling a Glacier ■
English Learners	❑ TE **Group Activity** No Glaciers?, p. 344 ❑ TE **Activity** Describing Glacier Formation, p. 345
Special Education Students	❑ TE **Universal Access** Classroom Glaciers, p. 344
Struggling Readers	❑ SE **Reading Strategy** Asking Questions, p. 344

Additional Resources

Holt Lab Generator CD-ROM

Search for any lab by topic, standard, difficulty level, or time. Edit any lab to fit your needs, or create your own labs. Use the Lab Materials QuickList software to customize your lab materials list. Lab datasheets are also available in Spanish on this CD-ROM.

Guided Reading Audio CD Program

The Guided Reading Audio CD Program provides a direct reading of the student text. This resource is helpful to auditory learners and struggling readers. This program is available in English and Spanish.

Key

SE Student Edition
TE Teacher's Edition
Chapter Resource File
Workbook
CD or CD-ROM
Transparency
Video
■ Also available in Spanish

All resources listed below are also available on the One-Stop Planner.

Focus on Earth Sciences: 6.2.a

Practice	Assess
❑ SE **Section Review,** p. 347 ❑ **Section Review** ■	❑ SE **Standards Checks,** pp. 344, 347 ❑ TE **Standards Focus,** p. 348 • Assess • Reteach • Re-Assess ❑ **Section Quiz** ■
❑ **Interactive Reader and Study Guide** ❑ **Vocabulary and Section Summary A** ■ ❑ **Reinforcement Worksheet**	
❑ **Interactive Reader and Study Guide** ❑ **Vocabulary and Section Summary A** ■ ❑ **Vocabulary and Section Summary B**	
❑ **Interactive Reader and Study Guide**	
❑ **Interactive Reader and Study Guide** ❑ **Directed Reading A** ■ ❑ **Directed Reading B**	❑ TE **Universal Access** Understanding Foreign Words, p. 345

Reviewing Prior Knowledge

Prepare students to learn about erosion and deposition by ice by asking them to speculate about landforms created by glaciers. See p. 344 of the Teacher's Edition.

MISCONCEPTION ALERT

Moving Ice Students may think that solids, such as ice, can not flow. To correct this misconception, see p. 345 in the Teacher's Edition.

 • This section should take approximately 1 day to complete.

Section 4 Erosion and Deposition by Mass Movement

Key Concept Gravity causes material to move downslope in a process called ***mass movement***.

		Teach
	Standards Course of Study	❑ **Bellringer Transparency** ❑ **PowerPoint® Resources** ❑ **E50** Angle of Repose ❑ **SE Quick Lab** Modeling a Landslide, p. 349 ❑ **Datasheet B for Quick Lab** Modeling a Landslide ■
Universal Access	Advanced Learners/GATE	❑ **Datasheet C for Quick Lab** Modeling a Landslide ■
Universal Access	Basic Learners	❑ **Datasheet A for Quick Lab** Modeling a Landslide ■
Universal Access	English Learners	❑ TE **Discussion** Erosion by Gravity, p. 348 ❑ TE **Activity** Demonstrating Mass Movement, p. 349 ❑ TE **Universal Access** Illustrating Mass Movements, p. 350
Universal Access	Special Education Students	
Universal Access	Struggling Readers	❑ **SE Reading Strategy** Graphic Organizer, p. 348 ❑ TE **Universal Access** Identifying Definitions in Text, p. 349

Additional Resources

Holt Lab Generator CD-ROM

Search for any lab by topic, standard, difficulty level, or time. Edit any lab to fit your needs, or create your own labs. Use the Lab Materials QuickList software to customize your lab materials list. Lab datasheets are also available in Spanish on this CD-ROM.

Guided Reading Audio CD Program

The Guided Reading Audio CD Program provides a direct reading of the student text. This resource is helpful to auditory learners and struggling readers. This program is available in English and Spanish.

Key

SE Student Edition
TE Teacher's Edition
Chapter Resource File
Workbook
CD or CD-ROM
Transparency
Video
■ Also available in Spanish

All resources listed below are also available on the One-Stop Planner.

Focus on Earth Sciences: 6.2.d
Math: Number Sense 6.2.1
English–Language Arts: Reading 6.2.4

Practice	Assess
❑ SE **Section Review,** p. 351 ❑ **Section Review** ■	❑ SE **Standards Checks,** pp. 349, 351 ❑ TE **Standards Focus,** p. 350 • Assess • Reteach • Re-Assess ❑ **Section Quiz** ■
❑ **Critical Thinking** ❑ **SciLinks Activity**	
❑ **Interactive Reader and Study Guide** ❑ **Vocabulary and Section Summary A** ■	
❑ **Interactive Reader and Study Guide** ❑ **Vocabulary and Section Summary A** ■ ❑ **Vocabulary and Section Summary B**	
❑ **Interactive Reader and Study Guide**	
❑ **Interactive Reader and Study Guide** ❑ **Directed Reading A** ■ ❑ **Directed Reading B**	

Reviewing Prior Knowledge

Prepare students to learn angle of repose by discussing piles of different materials. See p. 348 of the Teacher's Edition.

MISCONCEPTION ALERT

Preventing Mudslides Students may not think that lack of vegetation is a factor in mudslide occurrence. To correct this misconception, see p. 348 in the Teacher's Edition.

Math Support

Science and math go hand in hand. The Math Practice on p. 351 helps students practice math skills in a scientific context.

Pacing

- Chapter Lab, Review, and Assessment should take approximately 4 days to complete.

Wrapping Up

	Teach
Standards Course of Study	❑ SE **Model-Making Lab** Beach Erosion, pp. 352–353 ❑ **Datasheet B for Chapter Lab** Beach Erosion ■ ❑ **Standards Review Transparency** ■
Universal Access: Advanced Learners/GATE	❑ **Datasheet C for Chapter Lab** Beach Erosion ■
Universal Access: Basic Learners	❑ SE **Activity,** p. 361 ❑ **Datasheet A for Chapter Lab** Beach Erosion ■
Universal Access: English Learners	
Universal Access: Special Education Students	❑ TE **Universal Access** Picturing Tools, p. 354
Universal Access: Struggling Readers	❑ TE **Universal Access** Activating Prior Knowledge, p. 354

Additional Resources

SUPER SUMMARY

Have students review the major concepts in this chapter by using the Super Summary that includes the following:

- an outline of important points in the chapter
- flashcards for chapter vocabulary
- an interactive quiz

Go to **go.hrw.com**
Type in the keyword HY7ICES

Performance-Based Assessments

The Chapter Resource File for this chapter contains a hands-on activity that can be used to help assess student progress in a nontraditional format. In the Performance-Based Assessment for this chapter, students model the way in which glaciers move.

Key

SE Student Edition
TE Teacher's Edition
Chapter Resource File
Workbook
CD or CD-ROM
Transparency
Video
■ Also available in Spanish

All resources listed below are also available on the One-Stop Planner.

Focus on Earth Sciences: 6.2.a, 6.2.c, 6.3.a, 6.7.a, 6.7.b
Math: Number Sense 6.2.1
English–Language Arts: Writing 6.1.3

Practice	Assess
❑ SE **Science Skills Activity** Selecting Tools to Collect Data, p. 354 ❑ **Datasheet for Science Skills Activity** ■ ❑ **Concept Mapping Transparency** ❑ SE **Chapter Review,** pp. 356–357 ❑ **Chapter Review** ■	❑ **Chapter Test B** ■ ❑ SE **Standards Assessment,** pp. 358–359 ❑ **Standards Assessment** ❑ **Standards Review Workbook** ■
	❑ TE **Focus on Speaking,** p. 355 ❑ TE **Language Arts Activity,** p. 360 ❑ **Chapter Test C** ❑ **Brain Food Video Quiz**
❑ TE **Identifying Roots,** p. 355	❑ TE **Math Activity,** p. 360 ❑ **Chapter Test A** ❑ **Brain Food Video Quiz**
	❑ **Brain Food Video Quiz**
	❑ **Brain Food Video Quiz**
	❑ **Brain Food Video Quiz**

Holt Online Assessment

Post tests and quizzes to Holt Online Assessment, an assessment management tool. The system automatically grades the assessments, and you receive students' scores and information about which questions students missed. Holt Online Assessment is available through the Premier Online Edition of *Holt California Earth Science.*

Holt Anthology of Science Fiction

The Holt Anthology of Science Fiction includes thought-provoking stories that are relevant to science instruction. Enhance students' learning by asking them to read a story from the *Holt Anthology of Science Fiction* and to answer questions about what they have read.

Pacing

- This chapter should take approximately 10 days to complete.
- Getting Started should take approximately 1 day to complete.

Chapter 11 Rivers and Groundwater

Big Idea Topography is reshaped as water flows downhill in streams and rivers.

This chapter was designed to cover the California Grade 6 Science Standards about the influence of water on shaping Earth's surface (6.2.a, 6.2.b, 6.2.d, 6.4.a, and 6.6.b). It follows a chapter that introduced water as an agent of erosion and deposition. This chapter details erosion and deposition by streams and rivers and presents water as a resource.

After they have completed this chapter, students will begin a chapter about oceans.

Getting Started

		Teach
	Standards Course of Study	❑ SE **Explore Activity** The Sun and Water Cycle, p. 365 ❑ **Datasheet B for Explore Activity** The Sun and Water Cycle ■
Universal Access	Advanced Learners/GATE	❑ **Chapter Starter Transparency** ❑ **Datasheet C for Explore Activity** The Sun and Water Cycle ■
	Basic Learners	❑ SE **Improving Comprehension,** p. 362 ❑ **Chapter Starter Transparency** ❑ **Datasheet A for Explore Activity** The Sun and Water Cycle ■
	English Learners	❑ SE **Improving Comprehension,** p. 362 ❑ SE **Unpacking the Standards,** p. 363
	Special Education Students	❑ SE **Improving Comprehension,** p. 362 ❑ SE **Unpacking the Standards,** p. 363
	Struggling Readers	❑ SE **Improving Comprehension,** p. 362

Key

- **SE** Student Edition
- **TE** Teacher's Edition
- Chapter Resource File
- Workbook
- CD or CD-ROM
- Transparency
- Video
- ■ Also available in Spanish

All resources listed below are also available on the One-Stop Planner.

The California Science Standards listed below are covered in this chapter:

Focus on Earth Sciences

6.2.a Students know water running downhill is the dominant process in shaping the landscape, including California's landscape.

6.2.b Students know rivers and streams are dynamic systems that erode, transport sediment, change course, and flood their banks in natural and recurring patterns.

6.2.d Students know earthquakes, volcanic eruptions, landslides, and floods change human and wildlife habitats.

6.4.a Students know the sun is the major source of energy for phenomena on Earth's surface; it powers winds, ocean currents, and the water cycle.

6.6.b Students know different natural energy and material resources, including air, soil, rocks, minerals, petroleum, fresh water, wildlife, and forests, and know how to classify them as renewable or nonrenewable.

Investigation and Experimentation

6.7.a Develop a hypothesis.

6.7.d Communicate the steps and results from an investigation in written reports and oral presentations.

6.7.e Recognize whether evidence is consistent with a proposed explanation.

6.7.h Identify changes in natural phenomena over time without manipulating the phenomena (e.g., a tree limb, a grove of trees, a stream, a hillslope).

Practice	Assess
❑ **SE Organize Activity** Booklet, p. 364	❑ **Chapter Pretest**
❑ TE **Words with Multiple Meanings,** p. 363	
❑ TE **Using Other Graphic Organizers,** p. 362	

11 # Pacing • This section should take approximately 2 days to complete.

Section 1 The Active River

Key Concept Water running downhill is the dominant process in shaping the landscape.

		Teach
	Standards Course of Study	❑ **Bellringer Transparency** ❑ **PowerPoint® Resources** ❑ **E51** The Water Cycle ❑ SE **Quick Lab** River's Load, p. 371 ❑ **Datasheet B for Quick Lab** River's Load
Universal Access	Advanced Learners/GATE	❑ **Datasheet C for Quick Lab** River's Load ❑ **P15 Link to Physical Science** Changing the State of Water
	Basic Learners	❑ **Datasheet A for Quick Lab** River's Load
	English Learners	❑ TE **Universal Access** Watery Words, p. 368
	Special Education Students	❑ TE **Universal Access** Watery Words, p. 368 ❑ TE **Connection** Water Safety, p. 369 ❑ TE **Universal Access** River Load, p. 370
	Struggling Readers	❑ SE **Reading Strategy** Graphic Organizer, p. 366 ❑ TE **Reading Strategy** Prediction Guide, p. 370

Key

SE Student Edition
TE Teacher's Edition
Chapter Resource File
Workbook
CD or CD-ROM
Transparency
Video
■ Also available in Spanish

All resources listed below are also available on the One-Stop Planner.

Focus on Earth Sciences: 6.2.a, 6.2.b, 6.4.a

Practice	Assess
❑ SE **Section Review,** p. 373 ❑ **Section Review** ■	❑ SE **Standards Checks,** pp. 366, 367, 368, 369, 370, 371, 372 ❑ TE **Standards Focus,** p. 372 • Assess • Reteach • Re-Assess ❑ **Section Quiz** ■
❑ **SciLinks Activity**	❑ TE **Activity** Water Re-Cycle, p. 367 ❑ TE **Homework** World River Scrapbook, p. 368 ❑ TE **Connection Activity** Amazon Tours Web Page, p. 371
❑ **Interactive Reader and Study Guide** ❑ **Vocabulary and Section Summary A** ■	❑ TE **Activity** River Field Guide, p. 370
❑ TE **Demonstration** Fertile Sediment, p. 366 ❑ TE **Universal Access** Playing with Vocabulary, p. 366 ❑ TE **Group Activity** Mapping River Systems, p. 368 ❑ **Interactive Reader and Study Guide** ❑ **Vocabulary and Section Summary A** ■ ❑ **Vocabulary and Section Summary B**	
❑ **Interactive Reader and Study Guide**	
❑ TE **Universal Access** Using Context Clues, p. 367 ❑ TE **Universal Access** Understanding Words with Multiple Meanings, p. 369 ❑ **Interactive Reader and Study Guide** ❑ **Directed Reading A** ■ ❑ **Directed Reading B**	

Section 2 Stream and River Deposits

Key Concept Rivers and streams are dynamic systems that erode, transport sediment, change course, and flood their banks in natural and recurring patterns.

	Teach
Standards Course of Study	❑ **Bellringer Transparency** ❑ **PowerPoint® Resources** ❑ SE **Quick Lab** Make Your Own Lake, p. 376 ❑ **Datasheet B for Quick Lab** Make Your Own Lake ■
Universal Access: Advanced Learners/GATE	❑ **Datasheet C for Quick Lab** Make Your Own Lake ■
Universal Access: Basic Learners	❑ **Datasheet A for Quick Lab** Make Your Own Lake ■
Universal Access: English Learners	
Universal Access: Special Education Students	
Universal Access: Struggling Readers	❑ SE **Reading Strategy** Graphic Organizer, p. 374 ❑ TE **Universal Access** Recognizing Structural Patterns, p. 377

Additional Resources

Holt Lab Generator CD-ROM

Search for any lab by topic, standard, difficulty level, or time. Edit any lab to fit your needs, or create your own labs. Use the Lab Materials QuickList software to customize your lab materials list. Lab datasheets are also available in Spanish on this CD-ROM.

Guided Reading Audio CD Program

The Guided Reading Audio CD Program provides a direct reading of the student text. This resource is helpful to auditory learners and struggling readers. This program is available in English and Spanish.

Key

SE Student Edition
TE Teacher's Edition

Chapter Resource File
Workbook

CD or CD-ROM
Transparency

Video
■ Also available in Spanish

All resources listed below are also available on the One-Stop Planner.

Focus on Earth Sciences: 6.2.a, 6.2.b, 6.2.d
Math: Algebra and Functions 6.2.3

Practice	Assess
❑ SE **Section Review,** p. 377 ❑ **Section Review** ■	❑ SE **Standards Checks,** pp. 375, 377 ❑ TE **Standards Focus,** p. 376 • Assess • Reteach • Re-Assess ❑ **Section Quiz** ■
	❑ TE **Universal Access** Gold Rush, p. 374 ❑ **Long-Term Projects & Research Ideas** Canyon Controversy
❑ TE **Universal Access** Alluvial Fan or Delta?, p. 375 ❑ **Interactive Reader and Study Guide** ❑ **Vocabulary and Section Summary A** ■	
❑ TE **Activity** Debating Dams, p. 374 ❑ **Interactive Reader and Study Guide** ❑ **Vocabulary and Section Summary A** ■ ❑ **Vocabulary and Section Summary B**	❑ TE **Universal Access** Write About It!, p. 375
❑ **Interactive Reader and Study Guide**	
❑ **Interactive Reader and Study Guide** ❑ **Directed Reading A** ■ ❑ **Directed Reading B**	

Reviewing Prior Knowledge

Prepare students to learn how rivers erode, transport sediment, change course, and flood their banks by starting a discussion on standard 6.2.b. See page 374 of the Teacher's Edition.

MISCONCEPTION ALERT

Alluvial Fan Students may think that alluvial fans and deltas are the same feature. To correct this misconception, see page 375 in the Teacher's Edition.

Math Support

Science and math go hand in hand. The Math Skills question on p. 377 helps students practice math skills in a scientific context.

 • This section should take approximately 1.5 days to complete.

Section 3 Using Water Wisely

Key Concept Water resources can be endangered by pollution or overuse.

	Teach
Standards Course of Study	❑ **Bellringer Transparency** ❑ **PowerPoint® Resources** ❑ **E52** The Water Table and Wells ❑ **SE Quick Lab** How Much Water Can You Save?, p. 382 ❑ **Datasheet B for Quick Lab** How Much Water Can You Save? ■
Universal Access: Advanced Learners/GATE	❑ **Datasheet C for Quick Lab** How Much Water Can You Save? ■
Universal Access: Basic Learners	❑ **TE Discussion** Watching Water, p. 378 ❑ **Datasheet A for Quick Lab** How Much Water Can You Save? ■
Universal Access: English Learners	❑ **TE Wordwise** Word Connections, p. 379
Universal Access: Special Education Students	
Universal Access: Struggling Readers	❑ **SE Reading Strategy** Clarifying Concepts, p. 378

Additional Resources

Holt Lab Generator CD-ROM

Search for any lab by topic, standard, difficulty level, or time. Edit any lab to fit your needs, or create your own labs. Use the Lab Materials QuickList software to customize your lab materials list. Lab datasheets are also available in Spanish on this CD-ROM.

Guided Reading Audio CD Program

The Guided Reading Audio CD Program provides a direct reading of the student text. This resource is helpful to auditory learners and struggling readers. This program is available in English and Spanish.

Key

SE Student Edition
TE Teacher's Edition

Chapter Resource File
Workbook

CD or CD-ROM
Transparency

Video
■ Also available in Spanish

All resources listed below are also available on the One-Stop Planner.

Focus on Earth Sciences: 6.6.b
Math: Number Sense 6.1.3, 6.1.4
English–Language Arts: Reading 6.1.1

Practice	Assess
❑ SE **Section Review,** p. 383 ❑ **Section Review** ■	❑ SE **Standards Checks,** pp. 379, 381, 383 ❑ TE **Standards Focus,** p. 382 • Assess • Reteach • Re-Assess ❑ **Section Quiz** ■
❑ **Critical Thinking** ❑ TE **EcoLabs and Field Activities** The Frogs Are Off Course	❑ TE **Universal Access** Investigating Your Area, p. 379
❑ TE **Connection Activity** Finding Water, p. 379 ❑ **Interactive Reader and Study Guide** ❑ **Vocabulary and Section Summary A** ■ ❑ **Reinforcement Worksheet**	❑ TE **Homework** Presentation, p. 381
❑ TE **Using the Figure** Understanding Aquifers, p. 378 ❑ TE **Universal Access** Sources of Water Pollution, p. 380 ❑ **Interactive Reader and Study Guide** ❑ **Vocabulary and Section Summary A** ■ ❑ **Vocabulary and Section Summary B**	❑ TE **Group Activity** A Call for Conservation, p. 380
❑ TE **Universal Access** Drawing Water Problems, p. 381 ❑ **Interactive Reader and Study Guide**	
❑ **Interactive Reader and Study Guide** ❑ **Directed Reading A** ■ ❑ **Directed Reading B**	❑ TE **Universal Access** Drawing Conclusions, p. 379

Reviewing Prior Knowledge

Prepare students to learn about California water resources by starting a discussion on where the water that students drink comes from. See page 378 of the Teacher's Edition.

A Drop in the Ocean Students may think they cannot do anything to conserve water and prevent pollution. To correct this misconception, see page 379 in the Teacher's Edition.

Math Support

Science and math go hand in hand. The Math Practice on p. 380 helps students practice math skills in a scientific context.

Pacing

- Chapter Lab, Review, and Assessment should take approximately 4.5 days to complete.

Wrapping Up

	Teach
Standards Course of Study	❑ SE **Model-Making Lab** Carving a Stream, pp. 384–385 ❑ **Datasheet B for Chapter Lab** Carving a Stream ■ ❑ **Standards Review Transparency** ■
Universal Access: Advanced Learners/GATE	❑ **Datasheet C for Chapter Lab** Carving a Stream ■
Universal Access: Basic Learners	❑ TE **People in Science** Teaching Strategy, p. 393 ❑ **Datasheet A for Chapter Lab** Carving a Stream ■
Universal Access: English Learners	
Universal Access: Special Education Students	
Universal Access: Struggling Readers	

Additional Resources

SUPER SUMMARY

Have students review the major concepts in this chapter by using the Super Summary that includes the following:

- an outline of important points in the chapter
- flashcards for chapter vocabulary
- an interactive quiz

Go to **go.hrw.com**
Type in the keyword HY7DEPS

Performance-Based Assessments

The Chapter Resource File for this chapter contains a hands-on activity that can be used to help assess student progress in a nontraditional format. In the Performance-Based Assessment for this chapter, students build a model of an aquifer.

Key

SE Student Edition
TE Teacher's Edition
Chapter Resource File
Workbook
CD or CD-ROM
Transparency
Video
■ Also available in Spanish

All resources listed below are also available on the One-Stop Planner.

Focus on Earth Sciences: 6.2.a, 6.2.b, 6.2.d, 6.4.a, 6.6.b, 6.7.a, 6.7.d, 6.7.e, 6.7.h
Math: Number Sense 6.1.3, 6.1.4; Algebra and Functions 6.2.3
English–Language Arts: Writing 6.1.1, 6.1.3, 6.2.2, 6.2.5

Practice	Assess
❑ SE **Science Skills Activity** Identifying Changes over Time, p. 386 ❑ **Datasheet for Science Skills Activity** ■ ❑ **Concept Mapping Transparency** ❑ SE **Chapter Review,** pp. 388–389 ❑ **Chapter Review** ■	❑ **Chapter Test B** ■ ❑ SE **Standards Assessment,** pp. 390–391 ❑ **Standards Assessment** ❑ **Standards Review Workbook** ■
❑ SE **Social Studies Activity,** p. 392	❑ **Chapter Test C** ❑ **Brain Food Video Quiz**
❑ SE **Math Activity,** p. 393	❑ **Chapter Test A** ❑ **Brain Food Video Quiz**
❑ TE **Focus on Writing** Short Story, p. 387 ❑ TE **Identifying Suffixes** Vocabulary Suffixes, p. 387 ❑ SE **Language Arts Activity,** p. 392	❑ **Brain Food Video Quiz**
❑ TE **Universal Access** River Words, p. 386	❑ **Brain Food Video Quiz**
❑ TE **Universal Access** Visualizing Details, p. 386	❑ **Brain Food Video Quiz**

Holt Online Assessment

Post tests and quizzes to Holt Online Assessment, an assessment management tool. The system automatically grades the assessments, and you receive students' scores and information about which questions students missed. Holt Online Assessment is available through the Premier Online Edition of *Holt California Earth Science.*

Holt Anthology of Science Fiction

The Holt Anthology of Science Fiction includes thought-provoking stories that are relevant to science instruction. Enhance students' learning by asking them to read a story from the *Holt Anthology of Science Fiction* and to answer questions about what they have read.

Pacing

- This chapter should take approximately 10 days to complete.
- Getting Started should take approximately 1 day to complete.

Chapter 12 Exploring the Oceans

The Big Idea Oceans cover 71% of Earth's surface and contain natural resources that must be protected.

This chapter was designed to cover the California Grade 6 Science Standards about oceans (6.1.a, 6.1.d, 6.1.e, 6.3.c, 6.4.d, 6.6.a, 6.6.b, and 6.6.c). This chapter describes the ocean and its resources. The chapter also covers the discussion of environmental science and the wise use of resources and pollution prevention referred to in Category 1, Criterion 11, in the Criteria for Evaluating Instructional Materials in Science.

After they have completed this chapter, students will begin a chapter about the movement of ocean water.

Getting Started

		Teach
	Standards Course of Study	❑ **SE Explore Activity** Clean Up That Spill!, p. 399 ❑ **Datasheet B for Explore Activity** Clean Up That Spill! ■
Universal Access	Advanced Learners/GATE	❑ **Chapter Starter Transparency** ❑ **Datasheet C for Explore Activity** Clean Up That Spill! ■
	Basic Learners	❑ **SE Improving Comprehension,** p. 396 ❑ **Chapter Starter Transparency** ❑ **Datasheet A for Explore Activity** Clean Up That Spill! ■
	English Learners	❑ **SE Improving Comprehension,** p. 396 ❑ **SE Unpacking the Standards,** p. 397
	Special Education Students	❑ **SE Improving Comprehension,** p. 396 ❑ **SE Unpacking the Standards,** p. 397
	Struggling Readers	❑ **SE Improving Comprehension,** p. 396

Key

- **SE** Student Edition
- **TE** Teacher's Edition
- Chapter Resource File
- Workbook
- CD or CD-ROM
- Transparency
-  Video
- ■ Also available in Spanish

All resources listed below are also available on the One-Stop Planner.

The California Science Standards listed below are covered in this chapter:

Focus on Earth Sciences

6.1.a Students know evidence of plate tectonics is derived from the fit of the continents; the location of earthquakes, volcanoes, and midocean ridges; and the distribution of fossils, rock types, and ancient climatic zones.

6.1.d Students know that earthquakes are sudden motions along breaks in the crust called faults and that volcanoes and fissures are locations where magma reaches the surface.

6.1.e Students know major geologic events, such as earthquakes, volcanic eruptions, and mountain building, result from plate motions.

6.3.c Students know heat flows in solids by conduction (which involves no flow of matter) and in fluids by conduction and by convection (which involves flow of matter).

6.4.d Students know convection currents distribute heat in the atmosphere and oceans.

6.6.a Students know the utility of energy sources is determined by factors that are involved in converting these sources to useful forms and the consequences of the conversion process.

6.6.b Students know different natural energy and material resources, including air, soil, rocks, minerals, petroleum, fresh water, wildlife, and forests, and know how to classify them as renewable or nonrenewable.

6.6.c Students know the natural origin of the materials used to make common objects.

Investigation and Experimentation

6.7.a Develop a hypothesis.

6.7.e Recognize whether evidence is consistent with a proposed explanation.

6.7.g Interpret events by sequence and time from natural phenomena (e.g., the relative ages of rocks and intrusions).

Practice	Assess
❑ **SE Organize Activity** Four-Corner Fold, p. 398	❑ **Chapter Pretest**
❑ **TE Words with Multiple Meanings,** p. 397	
❑ **TE Using Other Graphic Organizers,** p. 396	

Pacing

• This section should take approximately 1 day to complete.

Section 1 Earth's Oceans

Key Concept The characteristics of ocean water, such as temperature and salinity, affect the circulation of the ocean.

		Teach
	Standards Course of Study	❑ **Bellringer Transparency** ❑ **PowerPoint® Resources** ❑ **E53** Divisions of the Global Ocean ❑ **E43** Temperature Zones in the Ocean ❑ **SE Quick Lab** Density Factors, p. 404 ❑ **Datasheet B for Quick Lab** Density Factors ■
Universal Access	Advanced Learners/GATE	❑ **P30 Link to Physical Science** Forming Sodium Chloride ❑ **Datasheet C for Quick Lab** Density Factors ■
	Basic Learners	❑ **TE Activity** Ocean Size, p. 400 ❑ **TE Universal Access** Testing Salinity, p. 401 ❑ **Datasheet A for Quick Lab** Density Factors ■
	English Learners	❑ **TE Using the Figure** Variable Salinity, p. 402 ❑ **TE Wordwise** Word Connection, p. 403
	Special Education Students	❑ **TE Universal Access** Global Perspective, p. 400
	Struggling Readers	❑ **SE Reading Strategy** Prediction Guide, p. 400 ❑ **TE Universal Access** Scanning Graphs, p. 403 ❑ **TE Wordwise** Word Connection, p. 403

Additional Resources

Holt Lab Generator CD-ROM

Search for any lab by topic, standard, difficulty level, or time. Edit any lab to fit your needs, or create your own labs. Use the Lab Materials QuickList software to customize your lab materials list. Lab datasheets are also available in Spanish on this CD-ROM.

Guided Reading Audio CD Program

The Guided Reading Audio CD Program provides a direct reading of the student text. This resource is helpful to auditory learners and struggling readers. This program is available in English and Spanish.

Key

SE Student Edition | TE Teacher's Edition | Chapter Resource File | Workbook | CD or CD-ROM | Transparency | Video | ■ Also available in Spanish

All resources listed below are also available on the One-Stop Planner.

Focus on Earth Sciences: 6.3.c, 6.4.d
Math: Number Sense 6.1.4

Practice	Assess
❑ SE **Section Review,** p. 405 ❑ **Section Review** ■	❑ SE **Standards Checks,** pp. 403, 405 ❑ TE **Standards Focus,** p. 404 • Assess • Reteach • Re-Assess ❑ **Section Quiz** ■
❑ TE **Debate** Long Rainy Days, p. 402 ❑ **SciLinks Activity**	
❑ TE **Activity** Diagramming Temperature Zones, p. 403 ❑ **Interactive Reader and Study Guide** ❑ **Vocabulary and Section Summary A** ■	❑ TE **Homework** Oceans and Weather, p. 403
❑ **Interactive Reader and Study Guide** ❑ **Vocabulary and Section Summary A** ■ ❑ **Vocabulary and Section Summary B**	
❑ **Interactive Reader and Study Guide**	
❑ **Interactive Reader and Study Guide** ❑ **Directed Reading A** ■ ❑ **Directed Reading B**	

Reviewing Prior Knowledge

Prepare students to learn about convection by having students review standard 6.3.c. See page 400 of the Teacher's Edition.

MISCONCEPTION ALERT

Sea or Ocean Students might confuse the term *sea* with the word *ocean.* To correct this misconception, see page 401 in the Teacher's Edition.

MISCONCEPTION ALERT

Less than Zero Because the deepest water in the Arctic Ocean is colder than 0°C, students may expect the water to be ice. To correct this misconception, see page 403 in the Teacher's Edition.

Pacing

• This section should take approximately 1 day to complete.

Section 2 The Ocean Floor

Key Concept Many different technologies have helped scientists study the topography of the ocean basins.

		Teach
	Standards Course of Study	❑ **Bellringer Transparency** ❑ **PowerPoint® Resources** ❑ **E55** Ocean Floor Features ❑ **SE Quick Lab** Seamounts, p. 410 ❑ **Datasheet B for Quick Lab** Seamounts ■
Universal Access	Advanced Learners/GATE	❑ **Datasheet C for Quick Lab** Seamounts ■
	Basic Learners	❑ **TE Activity** Research Invention, p. 406 ❑ **Datasheet A for Quick Lab** Seamounts ■
	English Learners	
	Special Education Students	❑ **TE Connection Activity** Illustrating the Ocean Floor, p. 409 ❑ **TE Universal Access** Knowing Names, p. 407
	Struggling Readers	❑ **SE Reading Strategy** Graphic Organizer, p. 406 ❑ **TE Connection Activity** Undersea Exploration, p. 407

Additional Resources

Holt Lab Generator CD-ROM

Search for any lab by topic, standard, difficulty level, or time. Edit any lab to fit your needs, or create your own labs. Use the Lab Materials QuickList software to customize your lab materials list. Lab datasheets are also available in Spanish on this CD-ROM.

Guided Reading Audio CD Program

The Guided Reading Audio CD Program provides a direct reading of the student text. This resource is helpful to auditory learners and struggling readers. This program is available in English and Spanish.

Key

SE Student Edition
TE Teacher's Edition

Chapter Resource File
Workbook

CD or CD-ROM
Transparency

Video
■ Also available in Spanish

All resources listed below are also available on the One-Stop Planner.

Focus on Earth Sciences: 6.1.a, 6.1.d, 6.1.e
Math: Algebra and Functions 6.1.1

Practice	Assess
❑ SE **Section Review,** p. 411 ❑ **Section Review** ■	❑ SE **Standards Checks,** pp. 409, 410 ❑ TE **Standards Focus,** p. 410 • Assess • Reteach • Re-Assess ❑ **Section Quiz** ■
❑ TE **Universal Access** Island Poster, p. 409 ❑ **Calculator-Based Labs** Ocean Floor Mapping	
❑ **Interactive Reader and Study Guide** ❑ **Vocabulary and Section Summary A** ■	
❑ TE **Universal Access** Fantasy Piloting, p. 407 ❑ TE **Using the Figure** Mid-Ocean Ridges, p. 409 ❑ **Interactive Reader and Study Guide** ❑ **Vocabulary and Section Summary A** ■ ❑ **Vocabulary and Section Summary B**	
❑ **Interactive Reader and Study Guide**	
❑ TE **Universal Access** Activating Prior Knowledge, p. 406 ❑ **Interactive Reader and Study Guide** ❑ **Directed Reading A** ■ ❑ **Directed Reading B**	

Reviewing Prior Knowledge

Prepare students to learn about the ocean floor by asking students what they know about how scientist have explored the ocean floor. See page 406 of the Teacher's Edition.

Varying Sea Levels Some students may think that sea level is the same worldwide. To correct this misconception, see page 408 in the Teacher's Edition.

Math Support

Science and math go hand in hand. The Math Skills item in the Section Review on p. 411 helps students practice math skills in a scientific context.

12 Pacing • This section should take approximately 2 days to complete.

Section 3 Resources from the Ocean

Key Concept The ocean is an important source of living and nonliving resources.

	Teach
Standards Course of Study	❑ **Bellringer Transparency** ❑ **PowerPoint® Resources** ❑ **E56** Using Tides to Generate Energy ❑ **SE Quick Lab** The Desalination Plant, p. 413 ❑ **Datasheet B for Quick Lab** The Desalination Plant ■
Universal Access: Advanced Learners/GATE	❑ **Datasheet C for Quick Lab** The Desalination Plant ■
Universal Access: Basic Learners	❑ **Datasheet A for Quick Lab** The Desalination Plant ■
Universal Access: English Learners	
Universal Access: Special Education Students	
Universal Access: Struggling Readers	❑ **SE Reading Strategy** Outlining, p. 412

Additional Resources

Holt Lab Generator CD-ROM

Search for any lab by topic, standard, difficulty level, or time. Edit any lab to fit your needs, or create your own labs. Use the Lab Materials QuickList software to customize your lab materials list. Lab datasheets are also available in Spanish on this CD-ROM.

Guided Reading Audio CD Program

The Guided Reading Audio CD Program provides a direct reading of the student text. This resource is helpful to auditory learners and struggling readers. This program is available in English and Spanish.

Key

SE Student Edition
TE Teacher's Edition
Chapter Resource File
Workbook
CD or CD-ROM
Transparency
Video
■ Also available in Spanish

All resources listed below are also available on the One-Stop Planner.

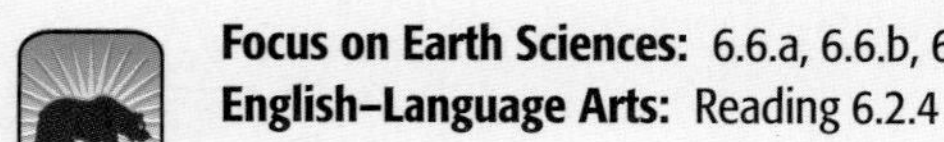

Focus on Earth Sciences: 6.6.a, 6.6.b, 6.6.c
English–Language Arts: Reading 6.2.4

Practice	Assess
❑ SE **Section Review,** p. 415 ❑ **Section Review** ■	❑ SE **Standards Checks,** pp. 412, 413, 414, 415 ❑ TE **Standards Focus,** p. 414 • Assess • Reteach • Re-Assess ❑ **Section Quiz** ■
❑ **Long-Term Projects & Research Ideas** Your Very Own Underwater Themepark	
❑ **Interactive Reader and Study Guide** ❑ **Vocabulary and Section Summary A** ■	
❑ TE **Group Activity** Brainstorming, p. 412 ❑ TE **Universal Access** Brainstorming Resources, p. 413 ❑ **Interactive Reader and Study Guide** ❑ **Vocabulary and Section Summary A** ■ ❑ **Vocabulary and Section Summary B**	
❑ TE **Universal Access** Renewable Resource List, p. 412 ❑ **Interactive Reader and Study Guide**	
❑ TE **Universal Access** Recognizing Context Clues, p. 412 ❑ TE **Reading Strategy** Anticipation Guide, p. 413 ❑ **Interactive Reader and Study Guide** ❑ **Directed Reading A** ■ ❑ **Directed Reading B**	

Reviewing Prior Knowledge

Prepare students to learn about ocean resources by having students read the red and blue heads in this section. See page 412 of the Teacher's Edition.

Ocean Phytoplankton Students may think that rain forests produce most of Earth's oxygen. To correct this misconception, see page 413 in the Teacher's Edition.

12 **Pacing** • This section should take approximately 1 day to complete.

Section 4 Ocean Pollution

Key Concept Activities on land and in the ocean contribute to ocean pollution.

	Teach
Standards Course of Study	❑ **Bellringer Transparency** ❑ **PowerPoint® Resources** ❑ **E57** Oil Spills ❑ SE **Quick Lab** Oily Feathers, p. 419 ❑ **Datasheet B for Quick Lab** Oily Feathers ■
Universal Access — Advanced Learners/GATE	❑ **Datasheet C for Quick Lab** Oily Feathers ■
Universal Access — Basic Learners	❑ TE **Connection Activity** Coastal Cleanup, p. 420 ❑ **Datasheet A for Quick Lab** Oily Feathers ■
Universal Access — English Learners	
Universal Access — Special Education Students	❑ TE **Universal Access** Flip Through, p. 416
Universal Access — Struggling Readers	❑ SE **Reading Strategy** Graphic Organizer, p. 416

Additional Resources

Holt Lab Generator CD-ROM

Search for any lab by topic, standard, difficulty level, or time. Edit any lab to fit your needs, or create your own labs. Use the Lab Materials QuickList software to customize your lab materials list. Lab datasheets are also available in Spanish on this CD-ROM.

Guided Reading Audio CD Program

The Guided Reading Audio CD Program provides a direct reading of the student text. This resource is helpful to auditory learners and struggling readers. This program is available in English and Spanish.

Key

SE Student Edition
TE Teacher's Edition
Chapter Resource File
Workbook
CD or CD-ROM
Transparency

Video
■ Also available in Spanish

All resources listed below are also available on the One-Stop Planner.

Focus on Earth Sciences: 6.6.a

Practice	Assess
❑ SE **Section Review,** p. 421 ❑ **Section Review** ■	❑ SE **Standards Check,** p. 419 ❑ TE **Standards Focus,** p. 420 • Assess • Reteach • Re-Assess ❑ **Section Quiz** ■
❑ TE **Universal Access** Nuclear Waste and the Ocean, p. 418 ❑ **Critical Thinking**	❑ TE **Homework** Investigating Your Area, p. 417 ❑ TE **Connection Activity** Global Ocean Pollution, p. 418
❑ **EcoLabs & Field Activities** Operation Oil-Spill Cleanup ❑ **Interactive Reader and Study Guide** ❑ **Vocabulary and Section Summary A** ■ ❑ **Reinforcement Worksheet**	
❑ TE **Universal Access** Ocean Pollution, p. 420 ❑ **Interactive Reader and Study Guide** ❑ **Vocabulary and Section Summary A** ■ ❑ **Vocabulary and Section Summary B**	❑ TE **Connection Activity** Ocean Pollution Awareness, p. 419
❑ TE **Activity** Poster Project, p. 416 ❑ **Interactive Reader and Study Guide**	
❑ TE **Universal Access** Drawing Conclusions, p. 417 ❑ **Interactive Reader and Study Guide** ❑ **Directed Reading A** ■ ❑ **Directed Reading B**	

Reviewing Prior Knowledge

Prepare students to learn ocean pollution by having students discuss the meaning of section vocabulary. See page 416 of the Teacher's Edition.

Natural Seepage Students may think all of the oil in ocean water comes from offshore oil-drilling activity. To correct this misconception, see page 419 in the Teacher's Edition.

Math Support

Science and math go hand in hand. The Math Practice on p. 421 helps students practice math skills in a scientific context.

Pacing

- Chapter Lab, Review, and Assessment should take approximately 4 days to complete.

Wrapping Up

		Teach
	Standards Course of Study	❑ SE **Skills Practice Lab** Investigating an Oil Spill, pp. 422–423 ❑ **Datasheet B for Chapter Lab** Investigating an Oil Spill ■ ❑ **Standards Review Transparency** ■
Universal Access	Advanced Learners/GATE	❑ **Datasheet C for Chapter Lab** Investigating an Oil Spill ■
	Basic Learners	❑ **Datasheet A for Chapter Lab** Investigating an Oil Spill ■
	English Learners	
	Special Education Students	❑ TE **Universal Access** Line Graph Review, p. 424
	Struggling Readers	❑ TE **Universal Access** Reading Graphs, p. 424

Additional Resources

SUPER SUMMARY

Have students review the major concepts in this chapter by using the Super Summary that includes the following:

- an outline of important points in the chapter
- flashcards for chapter vocabulary
- an interactive quiz

Go to **go.hrw.com**
Type in the keyword HY7OCES

Performance-Based Assessments

The Chapter Resource File for this chapter contains a hands-on activity that can be used to help assess student progress in a nontraditional format. In the Performance-Based Assessment for this chapter, students draw a map that shows a cross section of the ocean floor.

Key

- **SE** Student Edition
- **TE** Teacher's Edition
- Chapter Resource File
- Workbook
- CD or CD-ROM
- Transparency
- Video
- ■ Also available in Spanish

All resources listed below are also available on the One-Stop Planner.

Focus on Earth Sciences: 6.1.a, 6.1.d, 6.1.e, 6.3.c, 6.4.d, 6.6.a, 6.6.b, 6.6.c, 6.7.a, 6.7.e, 6.7.g
Math: Algebra and Functions 6.1.1; Number Sense 6.1.4
English–Language Arts: Writing 6.1.2, 6.1.3, 6.1.4

Practice	Assess
❑ **SE Science Skills Activity** Interpreting Events by Time, p. 424 ❑ **Datasheet for Science Skills Activity** ■ ❑ **Concept Mapping Transparency** ❑ **SE Chapter Review,** pp. 426–427 ❑ **Chapter Review** ■	❑ **Chapter Test B** ■ ❑ **SE Standards Assessment,** pp. 428–429 ❑ **Standards Assessment** ❑ **Standards Review Workbook** ■
	❑ **Chapter Test C** ❑ **Brain Food Video Quiz**
❑ **TE Activity** Winning Subs, p. 430 ❑ **SE Math Activity,** p. 430	❑ **Chapter Test A** ❑ **Brain Food Video Quiz**
❑ **TE Identifying Roots** Work in Pairs, p. 425 ❑ **TE Focus on Writing** Short Story, p. 425 ❑ **TE Language Arts Activity** p. 431	❑ **Brain Food Video Quiz**
❑ **TE Social Studies Activity,** p. 430	❑ **Brain Food Video Quiz**
❑ **TE Identifying Roots** Work in Pairs, p. 425	❑ **Brain Food Video Quiz**

Holt Online Assessment

Post tests and quizzes to Holt Online Assessment, an assessment management tool. The system automatically grades the assessments, and you receive students' scores and information about which questions students missed. Holt Online Assessment is available through the Premier Online Edition of *Holt California Earth Science.*

Holt Anthology of Science Fiction

The Holt Anthology of Science Fiction includes thought-provoking stories that are relevant to science instruction. Enhance students' learning by asking them to read a story from the *Holt Anthology of Science Fiction* and to answer questions about what they have read.

Pacing

- This chapter should take approximately 10 days to complete.
- Getting Started should take approximately 1 day to complete.

Chapter 13 The Movement of Ocean Water

The Big Idea The movement of ocean water is a major factor in energy transfer on Earth's surface.

This chapter was designed to cover the California Grade 6 Science Standards about the movement of ocean water and the transfer of energy through Earth's oceans (6.3.a, 6.3.c, 6.4.a, 6.4.d, and 6.4.e). It follows a chapter that introduced the ocean and its resources. This chapter describes how ocean currents and waves transfer energy on Earth's surface.

After they have completed this chapter, students will begin a chapter about the atmosphere.

Getting Started

		Teach
	Standards Course of Study	❑ **SE Explore Activity** The Ups and Downs of Convection, p. 435 ❑ **Datasheet B for Explore Activity** The Ups and Downs of Convection ■
Universal Access	Advanced Learners/GATE	❑ **Chapter Starter Transparency** ❑ **Datasheet C for Explore Activity** The Ups and Downs of Convection ■
Universal Access	Basic Learners	❑ **SE Improving Comprehension,** p. 432 ❑ **Chapter Starter Transparency** ❑ **Datasheet A for Explore Activity** The Ups and Downs of Convection ■
Universal Access	English Learners	❑ **SE Improving Comprehension,** p. 432 ❑ **SE Unpacking the Standards,** p. 433
Universal Access	Special Education Students	❑ **SE Improving Comprehension,** p. 432 ❑ **SE Unpacking the Standards,** p. 433
Universal Access	Struggling Readers	❑ **SE Improving Comprehension,** p. 432

Key

- **SE** Student Edition
- **TE** Teacher's Edition
- Chapter Resource File
- Workbook
- CD or CD-ROM
- Transparency
- Video
- Also available in Spanish

All resources listed below are also available on the One-Stop Planner.

The California Science Standards listed below are covered in this chapter:

Focus on Earth Sciences

6.3.a Students know energy can be carried from one place to another by heat flow or by waves, including water, light and sound waves, or by moving objects.

6.3.c Students know heat flows in solids by conduction (which involves no flow of matter) and in fluids by conduction and by convection (which involves flow of matter).

6.4.a Students know the sun is the major source of energy for phenomena on Earth's surface; it powers winds, ocean currents, and the water cycle.

6.4.d Students know convection currents distribute heat in the atmosphere and oceans.

6.4.e Students know differences in pressure, heat, air movement, and humidity result in changes of weather.

Investigation and Experimentation

6.7.a Develop a hypothesis.

6.7.e Recognize whether evidence is consistent with a proposed explanation.

Practice

- ❑ SE **Organize Activity** Three-Panel Flip Chart, p. 434
- ❑ TE **Words with Multiple Meanings,** p. 433
- ❑ TE **Using Other Graphic Organizers,** p. 432

Assess

- ❑ **Chapter Pretest**

Pacing

• This section should take approximately 2 days to complete.

Section 1 Currents

Key Concept The circulation of ocean water distributes water, heat, dissolved gases, and dissolved solids around Earth's surface.

		Teach
	Standards Course of Study	❑ **Bellringer Transparency** ❑ **PowerPoint® Resources** ❑ **E58** Global Winds ❑ **E59** Surface Ocean Currents ❑ **E60** How Ocean Water Becomes More Dense ❑ SE **Quick Lab** Creating Convection Currents, p. 441 ❑ **Datasheet B for Quick Lab** Creating Convection Currents ■ ❑ **E61** Circulation of Deep and Surface Currents ❑ **E62** Global Ocean Circulation
Universal Access	Advanced Learners/GATE	❑ TE **Discussion** Rivers and Surface Currents, p. 437 ❑ **Datasheet C for Quick Lab** Creating Convection Currents ■ ❑ **P64 Link to Physical Science** The Structure of the Sun
Universal Access	Basic Learners	❑ TE **Universal Access** Differences in Currents, p. 438 ❑ TE **Activity** The Coriolis Effect, p. 438 ❑ **Datasheet A for Quick Lab** Creating Convection Currents ■
Universal Access	English Learners	❑ TE **Universal Access** Demonstrating Currents, p. 438 ❑ TE **Activity** The Coriolis Effect, p. 438 ❑ TE **Using the Figure** Currents Between Cities, p. 438 ❑ TE **Wordwise** Convection, p. 439
Universal Access	Special Education Students	❑ TE **Universal Access** Wind Movements, p. 437 ❑ TE **Universal Access** Surface Acting, p. 439
Universal Access	Struggling Readers	❑ SE **Reading Strategy** Clarifying Concepts, p. 436 ❑ TE **Reading Strategy** Prediction Guide, p. 436 ❑ TE **Activity** The Coriolis Effect, p. 438 ❑ TE **Using the Figure** Currents Between Cities, p. 438

Key

SE Student Edition | Chapter Resource File | CD or CD-ROM | Video
TE Teacher's Edition | Workbook | Transparency | ■ Also available in Spanish

All resources listed below are also available on the One-Stop Planner.

Focus on Earth Sciences: 6.3.c, 6.4.a, 6.4.d, 6.7.a
Math: Number Sense 6.2.1; Algebra and Functions 6.2.3
English–Language Arts: Reading 6.1.1

Practice	Assess
❑ SE **Section Review,** p. 443 ❑ **Section Review** ■	❑ SE **Standards Checks,** pp. 437, 438, 439, 441, 442 ❑ TE **Standards Focus,** p. 442 • Assess • Reteach • Re-Assess ❑ **Section Quiz** ■
❑ **SciLinks Activity** ❑ TE **Connection to Math** Scientific Notation, p. 440	❑ TE **Connection Activity** Current Colonies, p. 437 ❑ TE **Universal Access** Incredible Voyages, p. 439 ❑ **Calculator-Based Labs** How Low Can You Go?
❑ **Interactive Reader and Study Guide** ❑ **Vocabulary and Section Summary A** ■ ❑ **Reinforcement Worksheet**	
❑ **Interactive Reader and Study Guide** ❑ **Vocabulary and Section Summary A** ■ ❑ **Vocabulary and Section Summary B** ❑ TE **Connection to Math** Scientific Notation, p. 440	
❑ **Interactive Reader and Study Guide**	
❑ TE **Universal Access** Creating a Comparison Chart, p. 440 ❑ **Interactive Reader and Study Guide** ❑ **Directed Reading A** ■ ❑ **Directed Reading B**	

Pacing • This section should take approximately 1 day to complete.

Section 2 Currents and Climate

Key Concept Ocean currents transport energy, affect climate and weather, and distribute nutrients.

	Teach
Standards Course of Study	❑ **Bellringer Transparency** ❑ **PowerPoint® Resources** ❑ **E63** Upwelling ❑ **SE Quick Lab** Warm Land, Cold Water, p. 446 ❑ **Datasheet B for Quick Lab** Warm Land, Cold Water ■
Universal Access: Advanced Learners/GATE	❑ **Datasheet C for Quick Lab** Warm Land, Cold Water ■
Universal Access: Basic Learners	❑ **Datasheet A for Quick Lab** Warm Land, Cold Water ■
Universal Access: English Learners	❑ TE **Discussion** The Effect of the Gulf Stream, p. 444 ❑ TE **Universal Access** Currents and Climate, p. 447
Universal Access: Special Education Students	❑ TE **Universal Access** El Niño and La Niña, p. 445
Universal Access: Struggling Readers	❑ SE **Reading Strategy** Graphic Organizer, p. 444 ❑ TE **Discussion** The Effect of the Gulf Stream, p. 444 ❑ TE **Universal Access** Using Foreign Words, p. 445

Additional Resources

Holt Lab Generator CD-ROM

Search for any lab by topic, standard, difficulty level, or time. Edit any lab to fit your needs, or create your own labs. Use the Lab Materials QuickList software to customize your lab materials list. Lab datasheets are also available in Spanish on this CD-ROM.

Guided Reading Audio CD Program

The Guided Reading Audio CD Program provides a direct reading of the student text. This resource is helpful to auditory learners and struggling readers. This program is available in English and Spanish.

Key

SE Student Edition
TE Teacher's Edition
Chapter Resource File
Workbook
CD or CD-ROM
Transparency

Video
■ Also available in Spanish

All resources listed below are also available on the One-Stop Planner.

Focus on Earth Sciences: 6.4.e
Math: Number Sense 6.1.4

Practice	Assess
❑ SE **Section Review,** p. 447 ❑ **Section Review** ■	❑ SE **Standards Checks,** pp. 445, 446 ❑ TE **Standards Focus,** p. 446 • Assess • Reteach • Re-Assess ❑ **Section Quiz** ■
❑ TE **Activity** Graphing Temperatures, p. 445	❑ TE **Research** El Niño Results, p. 445
❑ **Interactive Reader and Study Guide** ❑ **Vocabulary and Section Summary A** ■	
❑ TE **Activity** Graphing Temperatures, p. 445 ❑ TE **Universal Access** Currents and Climate, p. 447 ❑ **Interactive Reader and Study Guide** ❑ **Vocabulary and Section Summary A** ■ ❑ **Vocabulary and Section Summary B**	
❑ **Interactive Reader and Study Guide**	
❑ **Interactive Reader and Study Guide** ❑ **Directed Reading A** ■ ❑ **Directed Reading B**	

Reviewing Prior Knowledge

Prepare students to learn about how ocean currents affect climate by discussing standard 6.4.e. See page 444 of the Teacher's Edition.

Not All El Niños Are the Same
Students may think that every El Niño event causes the same changes in the atmosphere. To correct this misconception, see page 444 in the Teacher's Edition.

Math Support

Science and math go hand in hand.
The Math Skills item in the Section Review on page 447 helps students practice math skills in a scientific context.

Pacing • This section should take approximately 2 days to complete.

Section 3 Waves and Tides

Key Concept Energy is carried through the ocean by tides, which are caused by gravitational attraction between Earth, the moon, and the sun, and by waves.

		Teach
	Standards Course of Study	❑ **Bellringer Transparency** ❑ **PowerPoint® Resources** ❑ SE **Quick Lab** Making Waves, p. 449 ❑ **Datasheet B for Quick Lab** Making Waves ■ ❑ **E64** Determining Wave Period ❑ **E65** Why Waves Break ❑ **E66** Timing the Tides
Universal Access	Advanced Learners/GATE	❑ TE **Discussion** Offshore Breakers, p. 451 ❑ TE **Discussion** Tides and the Moon, p. 452 ❑ **Datasheet C for Quick Lab** Making Waves ■
Universal Access	Basic Learners	❑ TE **Demonstration** Making Waves, p. 448 ❑ TE **Universal Access** Wave in a Bottle, p. 448 ❑ **Datasheet A for Quick Lab** Making Waves ■
Universal Access	English Learners	❑ TE **Demonstration** Making Waves, p. 448 ❑ TE **Activity** Modeling Waves, p. 450 ❑ TE **Using the Figure** Breakers, p. 451 ❑ TE **Using the Figure** Earth's Rotation, p. 453 ❑ TE **Universal Access** Waves and Tides, p. 453
Universal Access	Special Education Students	❑ TE **Universal Access** Tsunami Study, p. 450 ❑ TE **Universal Access** Spring and Neap, p. 454
Universal Access	Struggling Readers	❑ SE **Reading Strategy** Graphic Organizer, p. 448 ❑ TE **Demonstration** Making Waves, p. 448 ❑ TE **Activity** Modeling Waves, p. 450 ❑ TE **Reading Strategy** Paired Summarizing, p. 451 ❑ TE **Using the Figure** Breakers, p. 451 ❑ TE **Universal Access** Heading Questions, p. 452

Key

SE Student Edition
TE Teacher's Edition
Chapter Resource File
Workbook
CD or CD-ROM
Transparency
Video
■ Also available in Spanish

All resources listed below are also available on the One-Stop Planner.

Focus on Earth Sciences: 6.3.a
Math: Number Sense 6.2.2; Algebra and Functions 6.2.3

Practice	Assess
❑ SE **Section Review,** p. 455 ❑ **Section Review** ■	❑ SE **Standards Checks,** pp. 448, 449, 450, 451 ❑ TE **Standards Focus,** p. 454 • Assess • Reteach • Re-Assess ❑ **Section Quiz** ■
❑ TE **Group Activity** Diagramming Waves, p. 451 ❑ TE **Homework** Graphing Tides, p. 453 ❑ **Critical Thinking**	❑ TE **Activity** Intertidal Zones, p. 452 ❑ TE **Homework** Surfing Report, p. 453 ❑ **Long-Term Projects & Research Ideas** An Ocean Commotion ❑ **Long-Term Projects & Research Ideas** Your Very Own Underwater Theme Park
❑ **Interactive Reader and Study Guide** ❑ **Vocabulary and Section Summary A** ■	
❑ TE **Group Activity** Diagramming Waves, p. 451 ❑ TE **Homework** Graphing Tides, p. 453 ❑ **Interactive Reader and Study Guide** ❑ **Vocabulary and Section Summary A** ■ ❑ **Vocabulary and Section Summary B**	
❑ **Interactive Reader and Study Guide**	
❑ TE **Group Activity** Diagramming Waves, p. 451 ❑ **Interactive Reader and Study Guide** ❑ **Directed Reading A** ■ ❑ **Directed Reading B**	

13 Pacing

- **Chapter Lab, Review, and Assessment should take approximately 4 days to complete.**

Wrapping Up

	Teach
Standards Course of Study	❑ SE **Skills Practice Lab** Modeling the Coriolis Effect, pp. 456–457 ❑ **Datasheet B for Chapter Lab** Modeling the Coriolis Effect ■ ❑ **Standards Review Transparency** ■
Universal Access — Advanced Learners/GATE	❑ **Datasheet C for Chapter Lab** Modeling the Coriolis Effect ■
Universal Access — Basic Learners	❑ **Datasheet A for Chapter Lab** Modeling the Coriolis Effect ■
Universal Access — English Learners	
Universal Access — Special Education Students	❑ TE **Universal Access** Preparation, p. 458
Universal Access — Struggling Readers	❑ TE **Universal Access** Understanding Comparisons, p. 458

Additional Resources

SUPER SUMMARY

Have students review the major concepts in this chapter by using the Super Summary that includes the following:

- an outline of important points in the chapter
- flashcards for chapter vocabulary
- an interactive quiz

Go to **go.hrw.com**
Type in the keyword HY7H2OS

Performance-Based Assessments

The Chapter Resource File for this chapter contains a hands-on activity that can be used to help assess student progress in a nontraditional format. In the Performance-Based Assessment for this chapter, students model the Coriolis effect and draw conclusions about how the Coriolis effect influences ocean currents.

Key

SE Student Edition
TE Teacher's Edition
Chapter Resource File
Workbook
CD or CD-ROM
Transparency
Video
■ Also available in Spanish

All resources listed below are also available on the One-Stop Planner.

Focus on Earth Sciences: 6.3.a, 6.3.c, 6.4.a, 6.4.d, 6.4.e, 6.7.a, 6.7.e
Math: Algebra and Functions 6.1.1; Statistics, Data, and Probability 6.1.1
English–Language Arts: Writing 6.2.1, 6.2.2

Practice	Assess
❑ SE **Science Skills Activity** Evaluating Explanations for Evidence, p. 458 ❑ **Datasheet for Science Skills Activity** ■ ❑ **Concept Mapping Transparency** ❑ SE **Chapter Review,** pp. 460–461 ❑ **Chapter Review** ■	❑ **Chapter Test B** ■ ❑ SE **Standards Assessment,** pp. 462–463 ❑ **Standards Assessment** ❑ **Standards Review Workbook** ■
❑ TE **Identifying Roots** Writing a Poem, Rap, or Song, p. 459 ❑ TE **Math Activity,** p. 464	❑ TE **Social Studies Activity,** p. 464 ❑ TE **Language Arts Activity,** p. 465 ❑ **Chapter Test C** ❑ **Brain Food Video Quiz**
	❑ **Chapter Test A** ❑ **Brain Food Video Quiz**
❑ TE **Identifying Roots** Writing a Poem, Rap, or Song, p. 459	❑ **Brain Food Video Quiz**
	❑ **Brain Food Video Quiz**
❑ TE **Focus on Speaking** Explaining the Standards, p. 459 ❑ TE **Activity** Learning from Accidents, p. 464	❑ **Brain Food Video Quiz**

Holt Online Assessment

Post tests and quizzes to Holt Online Assessment, an assessment management tool. The system automatically grades the assessments, and you receive students' scores and information about which questions students missed. Holt Online Assessment is available through the Premier Online Edition of *Holt California Earth Science.*

Holt Anthology of Science Fiction

The Holt Anthology of Science Fiction includes thought-provoking stories that are relevant to science instruction. Enhance students' learning by asking them to read a story from the *Holt Anthology of Science Fiction* and to answer questions about what they have read.

14

Pacing

- This chapter should take approximately 11 days to complete.
- Getting Started should take approximately 1 day to complete.

Chapter 14 The Atmosphere

The Big Idea Earth's atmosphere is a mixture of gases that absorbs solar energy and enables life on Earth.

This chapter was designed to cover the California Grade 6 Science Standards about the atmosphere (6.3.a, 6.3.c, 6.3.d, 6.4.a, 6.4.b, 6.4.d, 6.4.e, and 6.6.a). The chapter covers the discussion of environmental science and pollution prevention referred to in Category 1, Criterion 11, in the Criteria for Evaluating Instructional Materials in Science. This chapter introduces students to atmospheric heating and air movement within the atmosphere.

After they have completed this chapter, students will begin a chapter about weather and climate.

Getting Started

		Teach
	Standards Course of Study	❑ **SE Explore Activity** Sunlight and Temperature Change, p. 469 ❑ **Datasheet B for Explore Activity** Sunlight and Temperature Change ■
Universal Access	Advanced Learners/GATE	❑ **Chapter Starter Transparency** ❑ **Datasheet C for Explore Activity** Sunlight and Temperature Change ■
Universal Access	Basic Learners	❑ **SE Improving Comprehension,** p. 466 ❑ **Chapter Starter Transparency** ❑ **Datasheet A for Explore Activity** Sunlight and Temperature Change ■
Universal Access	English Learners	❑ **SE Improving Comprehension,** p. 466 ❑ **SE Unpacking the Standards,** p. 467
Universal Access	Special Education Students	❑ **SE Improving Comprehension,** p. 466 ❑ **SE Unpacking the Standards,** p. 467
Universal Access	Struggling Readers	❑ **SE Improving Comprehension,** p. 466

Key

SE Student Edition
TE Teacher's Edition
Chapter Resource File
Workbook
CD or CD-ROM
Transparency
Video
Also available in Spanish

All resources listed below are also available on the One-Stop Planner.

The California Science Standards listed below are covered in this chapter:

Focus on Earth Sciences

6.3.a Students know energy can be carried from one place to another by heat flow or by waves, including water, light and sound waves, or by moving objects.

6.3.c Students know heat flows in solids by conduction (which involves no flow of matter) and in fluids by conduction and by convection (which involves flow of matter).

6.3.d Students know heat energy is also transferred between objects by radiation (radiation can travel through space).

6.4.a Students know the sun is the major source of energy for phenomena on Earth's surface; it powers winds, ocean currents, and the water cycle.

6.4.b Students know solar energy reaches Earth through radiation, mostly in the form of visible light.

6.4.d Students know convection currents distribute heat in the atmosphere and oceans.

6.4.e Students know differences in pressure, heat, air movement, and humidity result in changes of weather.

6.6.a Students know the utility of energy sources is determined by factors that are involved in converting these sources to useful forms and the consequences of the conversion process.

Investigation and Experimentation

6.7.a Develop a hypothesis.

6.7.b Select and use appropriate tools and technology (including calculators, computers, balances, spring scales, microscopes, and binoculars) to perform tests, collect data, and display data.

6.7.d Communicate the steps and results from an investigation in written reports and oral presentations.

6.7.e Recognize whether evidence is consistent with a proposed explanation.

6.7.g Interpret events by sequence and time from natural phenomena (e.g., the relative ages of rocks and intrusions).

Practice	Assess
❑ **SE Organize Activity** Booklet, p. 468	❑ **Chapter Pretest**
❑ TE **Academic Vocabulary,** p. 467	
❑ TE **Using Other Graphic Organizers,** p. 466	

Pacing

• This section should take approximately 1 day to complete.

Section 1 Characteristics of the Atmosphere

Key Concept Earth's atmosphere absorbs solar energy and transports energy around Earth's surface.

		Teach
	Standards Course of Study	❑ **Bellringer Transparency** ❑ **PowerPoint® Resources** ❑ SE **Quick Lab** Modeling Air Pressure, p. 341 ❑ **E67** Layers of the Atmosphere ❑ **Datasheet B for Quick Lab** Modeling Air Pressure ■
Universal Access	Advanced Learners/GATE	❑ **Datasheet C for Quick Lab** Modeling Air Pressure ■
	Basic Learners	❑ **Datasheet A for Quick Lab** Modeling Air Pressure ■ ❑ TE **Universal Access** Modeling Air Pressure, p. 471
	English Learners	
	Special Education Students	
	Struggling Readers	❑ SE **Reading Strategy** Graphic Organizer, p. 470 ❑ TE **Universal Access** Clarifying Understanding, p. 470

Additional Resources

Holt Lab Generator CD-ROM

Search for any lab by topic, standard, difficulty level, or time. Edit any lab to fit your needs, or create your own labs. Use the Lab Materials QuickList software to customize your lab materials list. Lab datasheets are also available in Spanish on this CD-ROM.

Guided Reading Audio CD Program

The Guided Reading Audio CD Program provides a direct reading of the student text. This resource is helpful to auditory learners and struggling readers. This program is available in English and Spanish.

Key

SE Student Edition
TE Teacher's Edition
Chapter Resource File
Workbook
CD or CD-ROM
Transparency
Video
■ Also available in Spanish

All resources listed below are also available on the One-Stop Planner.

Focus on Earth Sciences: 6.4.b, 6.4.e
Math: Number Sense 6.2.1
English–Language Arts: Reading 6.2.4

Practice	Assess
❑ SE **Section Review,** p. 473 ❑ **Section Review** ■	❑ SE **Standards Checks,** pp. 471, 473 ❑ TE **Standards Focus,** p. 472 • Assess • Reteach • Re-Assess ❑ **Section Quiz** ■
❑ **SciLinks Activity**	
❑ **Interactive Reader and Study Guide** ❑ **Vocabulary and Section Summary A** ■ ❑ **Reinforcement Worksheet**	
❑ **Interactive Reader and Study Guide** ❑ **Vocabulary and Section Summary A** ■ ❑ **Vocabulary and Section Summary B** ❑ TE **Universal Access** Focusing on Layers, p. 472	
❑ **Interactive Reader and Study Guide** ❑ TE **Universal Access** Crossing Vocabulary Words, p. 472	
❑ **Interactive Reader and Study Guide** ❑ **Directed Reading A** ■ ❑ **Directed Reading B**	

Reviewing Prior Knowledge

Prepare students to learn about the composition of the atmosphere by discussing air and its effect on living things. See page 470 of the Teacher's Edition.

Steam Students may think that steam is a gas. To correct this misconception, see page 471 in the Teacher's Edition.

Math Support

Science and math go hand in hand. The Math Practice on p. 473 and the Math Skills item in the Section Review on p. 473 help students practice math skills in a scientific context.

14 Pacing • This section should take approximately 2 days to complete.

Section 2 Atmospheric Heating

Key Concept Heat in Earth's atmosphere is transferred by radiation, conduction, and convection.

		Teach
	Standards Course of Study	❑ **Bellringer Transparency** ❑ **PowerPoint® Resources** ❑ **E68** Scattering, Absorption, and Reflection ❑ **E69** Radiation from the Sun and the Electromagnetic Spectrum ❑ **E70** Radiation, Conduction, and Convection ❑ SE **Quick Lab** Modeling Air Movement by Convection, p. 236 ❑ **Datasheet B for Quick Lab** Modeling Air Movement by Convection ■
Universal Access	Advanced Learners/GATE	❑ **Datasheet C for Quick Lab** Modeling Air Movement by Convection ■
	Basic Learners	❑ **Datasheet A for Quick Lab** Modeling Air Movement by Convection ■ ❑ TE **Connection to Physical Science** Blue Atmosphere, p. 475
	English Learners	❑ TE **Demonstration** Popcorn, p. 474 ❑ TE **Using the Figure** Reflection or Absorption, p. 474
	Special Education Students	❑ TE **Universal Access** Balancing Act, p. 478
	Struggling Readers	❑ SE **Reading Strategy** Summarizing, p. 474 ❑ TE **Universal Access** Paired Summarizing, p. 475 ❑ TE **Reading Strategy** Reading Organizer, p. 475 ❑ TE **Universal Access** Activating Prior Knowledge, p. 478

Key

- **SE** Student Edition
- **TE** Teacher's Edition
- Chapter Resource File
- Workbook
- CD or CD-ROM
- Transparency
- Video
- ■ Also available in Spanish

All resources listed below are also available on the One-Stop Planner.

Focus on Earth Sciences: 6.3.a, 6.3.c, 6.3.d, 6.4.a, 6.4.b, 6.4.d, 6.7.e, 6.7.g
Math: Number Sense 6.2.1

Practice	Assess
❑ **SE Section Review,** p. 479 ❑ **Section Review** ■	❑ **SE Standards Checks,** pp. 474, 475, 476, 477, 478 ❑ **TE Standards Focus,** p. 478 • Assess • Reteach • Re-Assess ❑ **Section Quiz** ■
❑ **TE Connection to Environmental Science** Heat Island, p. 476 ❑ **Calculator-Based Lab** Heating of Land and Water ❑ **Calculator-Based Lab** The Greenhouse Effect	
❑ **TE Universal Access** Air Expansion, p. 477 ❑ **Interactive Reader and Study Guide** ❑ **Vocabulary and Section Summary A** ■ ❑ **EcoLabs & Field Activities** That Greenhouse Effect!	
❑ **TE Wordwise** Word Connections, p. 476 ❑ **TE Wordwise** Word Connections, p. 477 ❑ **Interactive Reader and Study Guide** ❑ **Vocabulary and Section Summary A** ■ ❑ **Vocabulary and Section Summary B**	
❑ **Interactive Reader and Study Guide**	
❑ **Interactive Reader and Study Guide** ❑ **Directed Reading A** ■ ❑ **Directed Reading B**	

Pacing • This section should take approximately 1 day to complete.

Section 3 Air Movement and Wind

Key Concept Global winds and local winds are produced by the uneven heating of Earth's surface.

		Teach
	Standards Course of Study	❑ **Bellringer Transparency** ❑ **PowerPoint® Resources** ❑ **E71** Pressure Belts ❑ SE **Quick Lab** Investigating the Coriolis Effect, p. 481 ❑ **E72** Sea and Land Breezes ❑ **E73** Valley Breezes and Mountain Breezes ❑ **Datasheet B for Quick Lab** Investigating the Coriolis Effect ■
Universal Access	Advanced Learners/GATE	❑ **Datasheet C for Quick Lab** Investigating the Coriolis Effect ■ ❑ TE **Universal Access** Understanding Spatial Patterns, p. 481
Universal Access	Basic Learners	❑ **Datasheet A for Quick Lab** Investigating the Coriolis Effect ■
Universal Access	English Learners	❑ TE **Demonstration** Air Movement, p. 480
Universal Access	Special Education Students	
Universal Access	Struggling Readers	❑ SE **Reading Strategy** Graphic Organizer, p. 480

Additional Resources

Holt Lab Generator CD-ROM

Search for any lab by topic, standard, difficulty level, or time. Edit any lab to fit your needs, or create your own labs. Use the Lab Materials QuickList software to customize your lab materials list. Lab datasheets are also available in Spanish on this CD-ROM.

Guided Reading Audio CD Program

The Guided Reading Audio CD Program provides a direct reading of the student text. This resource is helpful to auditory learners and struggling readers. This program is available in English and Spanish.

Key

SE Student Edition
TE Teacher's Edition
Chapter Resource File
Workbook
CD or CD-ROM
Transparency
Video
■ Also available in Spanish

All resources listed below are also available on the One-Stop Planner.

Focus on Earth Sciences: 6.4.a, 6.4.d, 6.4.e, 6.7.b
English–Language Arts: Reading 6.2.4

Practice	Assess
❑ SE **Section Review,** p. 483 ❑ **Section Review** ■	❑ SE **Standards Checks,** pp. 480, 482, 483 ❑ TE **Standards Focus,** p. 482 • Assess • Reteach • Re-Assess ❑ **Section Quiz** ■
❑ **Long-Term Projects & Research Ideas** A Breath of Fresh Ether?	
❑ TE **Activity** Cause and Effect, p. 481 ❑ **Interactive Reader and Study Guide** ❑ **Vocabulary and Section Summary A** ■	
❑ TE **Universal Access** Air Movement and Wind, p. 482 ❑ **Interactive Reader and Study Guide** ❑ **Vocabulary and Section Summary A** ■ ❑ **Vocabulary and Section Summary B**	
❑ **Interactive Reader and Study Guide**	
❑ **Interactive Reader and Study Guide** ❑ **Directed Reading A** ■ ❑ **Directed Reading B**	

Reviewing Prior Knowledge

Prepare students to learn about convection currents by reviewing the language of standard 6.4.d. See p. 480 of the Teacher's Edition.

Effects on Air Movement Students may think that pressure is the only factor that determines the direction of airflow. To correct this misconception, see p. 480 in the Teacher's Edition.

Pacing • This section should take approximately 2 days to complete.

Section 4 The Air We Breathe

Key Concept Air is an important natural resource that is affected by human activities.

	Teach
Standards Course of Study	❑ **Bellringer Transparency** ❑ **PowerPoint® Resources** ❑ **E74** Sources of Indoor Air Pollution ❑ **SE Quick Lab** Collecting Air-Pollution Particles, p. 487 ❑ **Datasheet B for Quick Lab** Collecting Air-Pollution Particles ■
Universal Access: Advanced Learners/GATE	❑ **L80 Link to Life Science** The Lungs ❑ **Datasheet C for Quick Lab** Collecting Air-Pollution Particles ■
Universal Access: Basic Learners	❑ TE **Connection to Physical Science** Incomplete Combustion, p. 485 ❑ **Datasheet A for Quick Lab** Collecting Air-Pollution Particles ■
Universal Access: English Learners	❑ TE **Demonstration** Smog in a Jar, p. 484 ❑ TE **Demonstration** Acid Rain, p. 487
Universal Access: Special Education Students	
Universal Access: Struggling Readers	❑ **SE Reading Strategy** Brainstorming, p. 484 ❑ TE **Universal Access** Evaluating Evidence, p. 487

Additional Resources

Holt Lab Generator CD-ROM

Search for any lab by topic, standard, difficulty level, or time. Edit any lab to fit your needs, or create your own labs. Use the Lab Materials QuickList software to customize your lab materials list. Lab datasheets are also available in Spanish on this CD-ROM.

Guided Reading Audio CD Program

The Guided Reading Audio CD Program provides a direct reading of the student text. This resource is helpful to auditory learners and struggling readers. This program is available in English and Spanish.

Key

SE Student Edition
TE Teacher's Edition
Chapter Resource File
Workbook
CD or CD-ROM
Transparency
Video
■ Also available in Spanish

All resources listed below are also available on the One-Stop Planner.

Focus on Earth Sciences: 6.6.a, 6.7.a

Practice	Assess
❑ SE **Section Review,** p. 491 ❑ **Section Review** ■	❑ SE **Standards Checks,** pp. 485, 486, 487, 488 ❑ TE **Standards Focus,** p. 490 • Assess • Reteach • Re-Assess ❑ **Section Quiz** ■
❑ TE **Connection to Real World** Local Air Pollution and Weather, p. 486 ❑ **Critical Thinking**	❑ TE **Universal Access** Smog Poetry, p. 485 ❑ TE **Universal Access** Acid Precipitation, p. 487 ❑ TE **Homework** Radon, p. 489
❑ TE **Activity** Classifying Pollutants, p. 485 ❑ **Interactive Reader and Study Guide** ❑ **Vocabulary and Section Summary A** ■ ❑ **Long-Term Projects & Research Ideas** There's Something in the Air	❑ SE **School-to-Home Activity** Air-Pollution Awareness, p. 490 ❑ TE **Homework** Ozone Holes, p. 488 ❑ TE **Connection to Health** Respiratory Diseases, p. 489
❑ TE **Universal Access** Primary and Secondary Pollutants, p. 484 ❑ **Interactive Reader and Study Guide** ❑ **Vocabulary and Section Summary A** ■ ❑ **Vocabulary and Section Summary B**	
❑ TE **Universal Access** Pollution Problems at Home, p. 486 ❑ **Interactive Reader and Study Guide**	
❑ **Interactive Reader and Study Guide** ❑ **Directed Reading A** ■ ❑ **Directed Reading B**	

Reviewing Prior Knowledge

Prepare students to learn about air pollution by having them brainstorm related words or phrases. See p. 484 of the Teacher's Edition.

Thin Layer of Ozone Students may think that the ozone layer is very thick. To correct this misconception, see p. 488 in the Teacher's Edition.

14

Pacing

- **Chapter Lab, Review, and Assessment should take approximately 4 days to complete.**

Wrapping Up

		Teach
	Standards Course of Study	❑ SE **Skills Practice Lab** Under Pressure!, pp. 492–493 ❑ **Datasheet B for Chapter Lab** Under Pressure! ■ ❑ **Standards Review Transparency** ■
Universal Access	Advanced Learners/GATE	❑ **Datasheet C for Chapter Lab** Under Pressure! ■
	Basic Learners	❑ **Datasheet A for Chapter Lab** Under Pressure! ■
	English Learners	
	Special Education Students	
	Struggling Readers	

Additional Resources

SUPER SUMMARY

Have students review the major concepts in this chapter by using the Super Summary that includes the following:

- an outline of important points in the chapter
- flashcards for chapter vocabulary
- an interactive quiz

Go to **go.hrw.com**
Type in the keyword HY7ATMS

Performance-Based Assessments

The Chapter Resource File for this chapter contains a hands-on activity that can be used to help assess student progress in a nontraditional format. In the Performance-Based Assessment for this chapter, students observe the effects of greenhouse gases on the atmosphere.

Key

- **SE** Student Edition
- **TE** Teacher's Edition
- Chapter Resource File
- Workbook
- CD or CD-ROM
- Transparency
- Video
- ■ Also available in Spanish

All resources listed below are also available on the One-Stop Planner.

Focus on Earth Sciences: 6.3.a, 6.3.c, 6.3.d, 6.4.b, 6.4.d, 6.4.e, 6.6.a, 6.7.a, 6.7.d, 6.7.e
English–Language Arts: Writing 6.1.3, Writing 6.2.5

Practice	Assess
❑ **SE Science Skills Activity** Preparing for an Oral Report, p. 494 ❑ **Datasheet for Science Skills Activity** ■ ❑ **Concept Mapping Transparency** ❑ **SE Chapter Review,** pp. 496–497 ❑ **Chapter Review** ■	❑ **Chapter Test B** ■ ❑ **SE Standards Assessment,** pp. 498–499 ❑ **Standards Assessment** ❑ **Standards Review Workbook** ■
❑ **TE Focus on Writing** Analyzing Summaries, p. 495	❑ **Chapter Test C** ❑ **Brain Food Video Quiz**
❑ **TE Identifying Root Words** Root of *Pollution,* p. 495 ❑ **SE Social Studies Activity,** p. 500 ❑ **SE Math Activity,** p. 500 ❑ **TE Activity** Researching Ozone, p. 500 ❑ **SE Language Arts Activity,** p. 501	❑ **Chapter Test A** ❑ **Brain Food Video Quiz**
❑ **TE Activity** Aerodynamic Design, p. 500	❑ **Brain Food Video Quiz**
	❑ **TE Universal Access** Accommodating Presentation Styles, p. 494 ❑ **Brain Food Video Quiz**
	❑ **TE Universal Access** Organizing Information, p. 494 ❑ **Brain Food Video Quiz**

Holt Online Assessment

Post tests and quizzes to Holt Online Assessment, an assessment management tool. The system automatically grades the assessments, and you receive students' scores and information about which questions students missed. Holt Online Assessment is available through the Premier Online Edition of *Holt California Earth Science.*

Holt Anthology of Science Fiction

The Holt Anthology of Science Fiction includes thought-provoking stories that are relevant to science instruction. Enhance students' learning by asking them to read a story from the *Holt Anthology of Science Fiction* and to answer questions about what they have read.

15 Pacing

- This chapter should take approximately 14 days to complete.
- Getting Started should take approximately 1 day to complete.

Chapter 15 Weather and Climate

The Big Idea Differences in pressure, heat, air movement, and humidity result in changes of weather and thus affect climate.

This chapter was designed to cover the California Grade 6 Science Standards about weather and climate (6.2.d, 6.4.a, 6.4.b, 6.4.d, 6.4.e, and 6.6.a). It follows a chapter that introduced atmospheric heating and air movement in the atmosphere. This chapter describes how air pressure, air movement, humidity, and solar radiation influence Earth's weather and climate.

After they have completed this chapter, students will begin a unit that focuses on ecology.

Getting Started

		Teach
	Standards Course of Study	❑ **SE Explore Activity** The Meeting of Air Masses, p. 505 ❑ **Datasheet B for Explore Activity** The Meeting of Air Masses ■
Universal Access	Advanced Learners/GATE	❑ **Chapter Starter Transparency** ❑ **Datasheet C for Explore Activity** The Meeting of Air Masses ■
Universal Access	Basic Learners	❑ **SE Improving Comprehension,** p. 502 ❑ **Chapter Starter Transparency** ❑ **Datasheet A for Explore Activity** The Meeting of Air Masses ■
Universal Access	English Learners	❑ **SE Improving Comprehension,** p. 502 ❑ **SE Unpacking the Standards,** p. 503
Universal Access	Special Education Students	❑ **SE Improving Comprehension,** p. 502 ❑ **SE Unpacking the Standards,** p. 503
Universal Access	Struggling Readers	❑ **SE Improving Comprehension,** p. 502

Key

SE Student Edition
TE Teacher's Edition
Chapter Resource File
Workbook
CD or CD-ROM
Transparency
Video
■ Also available in Spanish

All resources listed below are also available on the One-Stop Planner.

The California Science Standards listed below are covered in this chapter:

Focus on Earth Sciences

6.2.d Students know earthquakes, volcanic eruptions, landslides, and floods change human and wildlife habitats.

6.4.a Students know the sun is the major source of energy for phenomena on Earth's surface; it powers winds, ocean currents, and the water cycle.

6.4.b Students know solar energy reaches Earth through radiation, mostly in the form of visible light.

6.4.d Students know convection currents distribute heat in the atmosphere and oceans.

6.4.e Students know differences in pressure, heat, air movement, and humidity result in changes of weather.

6.6.a Students know the utility of energy sources is determined by factors that are involved in converting these sources to useful forms and the consequences of the conversion process.

Investigation and Experimentation

6.7.b Select and use appropriate tools and technology (including calculators, computers, balances, spring scales, microscopes, and binoculars) to perform tests, collect data, and display data.

Practice	Assess
❑ SE **Organize Activity** Double Door, p. 504	❑ **Chapter Pretest**
❑ TE **Words with Multiple Meanings,** p. 503	
❑ TE **Using Other Graphic Organizers,** p. 502	

Pacing

• This section should take approximately 2 days to complete.

Section 1 Water in the Air

Key Concept The sun's energy drives the water cycle and causes differences in humidity that may lead to precipitation.

		Teach
	Standards Course of Study	❑ **Bellringer Transparency** ❑ **PowerPoint® Resources** ❑ **E75** Cloud Types based on Shape and Altitude ❑ **SE Quick Lab** Reaching the Dew Point, p. 509 ❑ **Datasheet B for Quick Lab** Reaching the Dew Point ■
Universal Access	Advanced Learners/GATE	❑ **Datasheet C for Quick Lab** Reaching the Dew Point ■ ❑ **L36 Link to Life Science** Geologic Time Scale
Universal Access	Basic Learners	❑ TE **Discussion** What Dew You Think?, p. 509 ❑ **Datasheet A for Quick Lab** Reaching the Dew Point ■
Universal Access	English Learners	
Universal Access	Special Education Students	❑ TE **Universal Access** Cloud Types, p. 510 ❑ TE **Discussion** Water in the Air, p. 508
Universal Access	Struggling Readers	❑ **SE Reading Strategy** Summarizing, p. 506 ❑ TE **Universal Access** Drawing Conclusions, p. 507

Additional Resources

Holt Lab Generator CD-ROM

Search for any lab by topic, standard, difficulty level, or time. Edit any lab to fit your needs, or create your own labs. Use the Lab Materials QuickList software to customize your lab materials list. Lab datasheets are also available in Spanish on this CD-ROM.

Guided Reading Audio CD Program

The Guided Reading Audio CD Program provides a direct reading of the student text. This resource is helpful to auditory learners and struggling readers. This program is available in English and Spanish.

Key

SE Student Edition
TE Teacher's Edition
Chapter Resource File
Workbook
CD or CD-ROM
Transparency
Video
■ Also available in Spanish

All resources listed below are also available on the One-Stop Planner.

Focus on Earth Sciences: 6.4.a, 6.4.e
Math: Algebra and Functions 6.1.1

Practice	Assess
❑ SE **Section Review,** p. 511 ❑ **Section Review** ■	❑ SE **Standards Checks,** pp. 506, 509, 510, 511 ❑ TE **Standards Focus,** p. 510 • Assess • Reteach • Re-Assess ❑ **Section Quiz** ■
❑ TE **Universal Access** Hair Hygrometer, p. 508 ❑ **SciLinks Activity**	
❑ TE **Universal Access** Naming Clouds, p. 510 ❑ **Interactive Reader and Study Guide** ❑ **Vocabulary and Section Summary A** ■ ❑ **Calculator-Based Labs** Relative Humidity	
❑ TE **Group Activity** Air Molecules, p. 506 ❑ **Interactive Reader and Study Guide** ❑ **Vocabulary and Section Summary A** ■ ❑ **Vocabulary and Section Summary B**	❑ TE **Universal Access** The Water Cycle, p. 506
❑ **Interactive Reader and Study Guide**	
❑ **Interactive Reader and Study Guide** ❑ **Directed Reading A** ■ ❑ **Directed Reading B**	

Reviewing Prior Knowledge

Prepare students to learn new words by reading each vocabulary word from the section. Then, ask the students what they think each word means. See page 506 of the Teacher's Edition.

Contrails Students may think that it is smoke that trails behind an airplane in the sky. To correct this misconception, see page 507 in the Teacher's Edition.

Math Support

Science and math go hand in hand. The Math Practice on p. 507 helps students practice math skills in a scientific context.

Pacing • This section should take approximately 2 days to complete.

Section 2 Fronts and Weather

Key Concept Weather results from the movement of air masses that differ in temperature and humidity.

		Teach
	Standards Course of Study	❑ **Bellringer Transparency** ❑ **PowerPoint® Resources** ❑ **E76** The Four Main Types of Fronts ❑ **SE Quick Lab** Modeling a Front, p. 514 ❑ **Datasheet B for Quick Lab** Modeling a Front ■ ❑ **E77** Hurricane
Universal Access	Advanced Learners/GATE	❑ **Datasheet C for Quick Lab** Modeling a Front ■
Universal Access	Basic Learners	❑ TE **Discussion** Qualities of Air, p. 512 ❑ TE **Demonstration** Modeling Thunder, p. 515 ❑ TE **Universal Access** Model Tornado, p. 516 ❑ **Datasheet A for Quick Lab** Modeling a Front ■
Universal Access	English Learners	❑ TE **Wordwise** Word Connection, p. 514
Universal Access	Special Education Students	
Universal Access	Struggling Readers	❑ **SE Reading Strategy** Graphic Organizer, p. 512 ❑ TE **Universal Access** Understanding Words with Multiple Meanings, p. 512

Additional Resources

Holt Lab Generator CD-ROM

Search for any lab by topic, standard, difficulty level, or time. Edit any lab to fit your needs, or create your own labs. Use the Lab Materials QuickList software to customize your lab materials list. Lab datasheets are also available in Spanish on this CD-ROM.

Guided Reading Audio CD Program

The Guided Reading Audio CD Program provides a direct reading of the student text. This resource is helpful to auditory learners and struggling readers. This program is available in English and Spanish.

Key

SE Student Edition
TE Teacher's Edition
Chapter Resource File
Workbook
CD or CD-ROM
Transparency

Video
■ Also available in Spanish

All resources listed below are also available on the One-Stop Planner.

Focus on Earth Sciences: 6.2.d, 6.4.a, 6.4.e

Practice	Assess
❑ SE **Section Review,** p. 519 ❑ **Section Review** ■	❑ SE **Standards Checks,** pp. 513, 514, 517, 518 ❑ TE **Standards Focus,** p. 518 • Assess • Reteach • Re-Assess ❑ **Section Quiz** ■
❑ TE **Activity** Using Maps, p. 513 ❑ **EcoLabs & Field Activities** Rain Maker or Rain Faker? ❑ **Long-Term Projects & Research Ideas** A Storm on the Horizon	❑ TE **Activity** Weather and Energy, p. 516 ❑ TE **Universal Access** Disaster Plan, p. 518
❑ **Interactive Reader and Study Guide** ❑ **Vocabulary and Section Summary A** ■ ❑ **Reinforcement Worksheet**	
❑ TE **Universal Access** Studying Fronts, p. 513 ❑ TE **Using the Figure** Tornado, p. 516 ❑ **Interactive Reader and Study Guide** ❑ **Vocabulary and Section Summary A** ■ ❑ **Vocabulary and Section Summary B**	❑ TE **Group Activity** Hurricane Newscast, p. 517
❑ TE **Universal Access** Checking Weather Reports, p. 512 ❑ **Interactive Reader and Study Guide**	
❑ **Interactive Reader and Study Guide** ❑ **Directed Reading A** ■ ❑ **Directed Reading B**	

Reviewing Prior Knowledge

Prepare students to learn about fronts by asking them what they think a front is. See page 512 of the Teacher's Edition.

MISCONCEPTION ALERT

Humid Is Lighter Than Dry Is Students may think that humid air is heavier than dry air is. To correct this misconception, see page 514 in the Teacher's Edition.

MISCONCEPTION ALERT

"The Same Place Twice" Students may think that lightning never strikes twice in the same place. To correct this misconception, see page 515 in the Teacher's Edition.

Pacing • This section should take approximately 2 days to complete.

Section 3 What Is Climate?

Key Concept Earth's climate zones are caused by the distribution of heat around Earth's surface by wind and ocean currents.

	Teach
Standards Course of Study	❑ **Bellringer Transparency** ❑ **PowerPoint® Resources** ❑ **E78** The Seasons ❑ **E79** Global Winds ❑ **SE Quick Lab** A Cool Breeze, p. 525 ❑ **Datasheet B for Quick Lab** A Cool Breeze ■ ❑ **E80** Earth's Climate
Universal Access: Advanced Learners/GATE	❑ **Datasheet C for Quick Lab** A Cool Breeze ■
Universal Access: Basic Learners	❑ **TE Discussion** Latitude and Climate, p. 520 ❑ **Datasheet A for Quick Lab** A Cool Breeze ■
Universal Access: English Learners	❑ **TE Using the Figure** Comparing Climates, p. 520 ❑ **TE Universal Access** Demonstrating Climate, p. 520 ❑ **TE Group Activity** Modeling the Sun and Earth, p. 522 ❑ **TE Using the Figure** Prevailing Winds, p. 523 ❑ **TE Using the Figure** Wet Winds, p. 524
Universal Access: Special Education Students	❑ **TE Universal Access** Move Like the Wind, p. 523
Universal Access: Struggling Readers	❑ **SE Reading Strategy** Graphic Organizer, p. 520

Key

- SE Student Edition
- TE Teacher's Edition
- Chapter Resource File
- Workbook
- CD or CD-ROM
- Transparency
- Video
- ■ Also available in Spanish

All resources listed below are also available on the One-Stop Planner.

Focus on Earth Sciences: 6.4.b, 6.4.d, 6.4.e

Practice	Assess
❑ SE **Section Review,** p. 527 ❑ **Section Review** ■	❑ SE **Standards Checks,** pp. 520, 522, 523 ❑ TE **Standards Focus,** p. 526 • Assess • Reteach • Re-Assess ❑ **Section Quiz** ■
❑ TE **Universal Access** Using Maps, p. 522 ❑ **Calculator-Based Labs** What Causes the Seasons?	
❑ **Interactive Reader and Study Guide** ❑ **Vocabulary and Section Summary A** ■	
❑ TE **Group Activity** Latitude and Temperature, p. 521 ❑ **Interactive Reader and Study Guide** ❑ **Vocabulary and Section Summary A** ■ ❑ **Vocabulary and Section Summary B**	
❑ **Interactive Reader and Study Guide**	
❑ TE **Universal Access** Identifying Structural Patterns, p. 521 ❑ TE **Universal Access** Retelling Ideas, p. 524 ❑ TE **Reading Strategy** Paired Summarizing, p. 525 ❑ **Interactive Reader and Study Guide** ❑ **Directed Reading A** ■ ❑ **Directed Reading B**	

 • This section should take approximately 2 days to complete.

Section 4 Changes in Climate

Key Concept Climate changes may be caused by natural factors and by human activities and can affect human and wildlife habitats.

		Teach
	Standards Course of Study	❑ **Bellringer Transparency** ❑ **PowerPoint® Resources** ❑ SE **Quick Lab** Hot Stuff, p. 532 ❑ **Datasheet B for Quick Lab** Hot Stuff ■
Universal Access	Advanced Learners/GATE	❑ **Datasheet C for Quick Lab** Hot Stuff ■
	Basic Learners	❑ TE **Demonstration** The Greenhouse Effect, p. 528 ❑ **Datasheet A for Quick Lab** Hot Stuff ■
	English Learners	
	Special Education Students	
	Struggling Readers	❑ SE **Reading Strategy** Clarifying Concepts, p. 528 ❑ TE **Universal Access** Using the Glossary, p. 528

Additional Resources

Holt Lab Generator CD-ROM

Search for any lab by topic, standard, difficulty level, or time. Edit any lab to fit your needs, or create your own labs. Use the Lab Materials QuickList software to customize your lab materials list. Lab datasheets are also available in Spanish on this CD-ROM.

Guided Reading Audio CD Program

The Guided Reading Audio CD Program provides a direct reading of the student text. This resource is helpful to auditory learners and struggling readers. This program is available in English and Spanish.

Key

SE Student Edition
TE Teacher's Edition
Chapter Resource File
Workbook
CD or CD-ROM
Transparency
Video
■ Also available in Spanish

All resources listed below are also available on the One-Stop Planner.

Focus on Earth Sciences: 6.2.d, 6.4.a, 6.4.b, 6.6.a
Math Algebra and Functions: 6.1.1
English–Language Arts: 6.1.2

Practice	Assess
❑ SE **Section Review,** p. 533 ❑ **Section Review** ■	❑ SE **Standards Checks,** pp. 529, 530, 531, 532 ❑ TE **Standards Focus,** p 532 • Assess • Reteach • Re-Assess ❑ **Section Quiz** ■
❑ **Critical Thinking** ❑ **Calculator-Based Labs** Sun-Starved in Fairbanks	❑ TE **Activity** Ancient Climates, p. 529 ❑ TE **Universal Access** Global Warming, p. 531
❑ **Interactive Reader and Study Guide** ❑ **Vocabulary and Section Summary A** ■	❑ TE **Activity** Volcanic Eruptions, p. 531 ❑ TE **Universal Access** Poster Project, p. 531
❑ TE **Universal Access** Changes in Climate, p. 529 ❑ **Interactive Reader and Study Guide** ❑ **Vocabulary and Section Summary A** ■ ❑ **Vocabulary and Section Summary B**	
❑ **Interactive Reader and Study Guide**	
❑ **Interactive Reader and Study Guide** ❑ **Directed Reading A** ■ ❑ **Directed Reading B**	

Reviewing Prior Knowledge

Prepare students to learn about climate change by having students read the red and blue heads in the section aloud. See page 528 of the Teacher's Edition.

Ice Ages and Glacial Periods Students may be confused about the difference between an ice age and a glacial period. To correct this misconception, see page 529 in the Teacher's Edition.

Math Support

Science and math go hand in hand. The Math Skills item in the Section Review on p. 533 helps students practice math skills in a scientific context.

Pacing

- **Chapter Lab, Review, and Assessment should take approximately 5 days to complete.**

Wrapping Up

	Teach
Standards Course of Study	❑ **SE Skills Practice Lab** Convection Currents, pp. 534–535 ❑ **Datasheet B for Chapter Lab** Convection Currents ■ ❑ **Standards Review Transparency** ■
Universal Access: Advanced Learners/GATE	❑ **Datasheet C for Chapter Lab** Convection Currents ■
Universal Access: Basic Learners	❑ **Datasheet A for Chapter Lab** Convection Currents ■
Universal Access: English Learners	
Universal Access: Special Education Students	
Universal Access: Struggling Readers	

Additional Resources

SUPER SUMMARY

Have students review the major concepts in this chapter by using the Super Summary that includes the following:

- an outline of important points in the chapter
- flashcards for chapter vocabulary
- an interactive quiz

Go to **go.hrw.com**
Type in the keyword HY7CLMS

Performance-Based Assessments

The Chapter Resource File for this chapter contains a hands-on activity that can be used to help assess student progress in a nontraditional format. In the Performance-Based Assessment for this chapter, students will make a model of the ocean and observe how currents affect climate.

Key
SE Student Edition
TE Teacher's Edition
Chapter Resource File
Workbook
CD or CD-ROM
Transparency
Video
■ Also available in Spanish

All resources listed below are also available on the One-Stop Planner.

Focus on Earth Sciences: 6.2.d, 6.4.a, 6.4.b, 6.4.d, 6.4.e, 6.6.a, 6.7.b
Math: Algebra and Functions 6.1.1
English–Language Arts: Reading 6.1.1; Writing 6.1.2, 6.1.3, 6.2.1, 6.2.2

Practice	Assess
❑ SE **Science Skills Activity** Collecting Weather Data, p. 536 ❑ **Datasheet for Science Skills Activity** ■ ❑ **Concept Mapping Transparency** ❑ SE **Chapter Review,** pp. 538–539 ❑ **Chapter Review** ■	❑ **Chapter Test B** ■ ❑ SE **Standards Assessment,** pp. 540–541 ❑ **Standards Assessment** ❑ **Standards Review Workbook** ■
	❑ TE **Connection Activity** Forensic Science, p. 543 ❑ **Chapter Test C** ❑ **Brain Food Video Quiz**
❑ SE **Math Activity,** p. 542 ❑ SE **Social Studies Activity,** p. 543	❑ TE **Homework** Local Weather Forecast, p. 543 ❑ **Chapter Test A** ❑ **Brain Food Video Quiz**
❑ TE **Identifying Suffixes** Group Activity, p. 537 ❑ TE **Focus on Writing** Connecting Concepts, p. 537 ❑ SE **Language Arts Activity,** p. 542	❑ **Brain Food Video Quiz**
❑ TE **Universal Access** Inferring, p. 536 ❑ SE **Language Arts Activity,** p. 542	❑ **Brain Food Video Quiz**
	❑ **Brain Food Video Quiz**

Holt Online Assessment

Post tests and quizzes to Holt Online Assessment, an assessment management tool. The system automatically grades the assessments, and you receive students' scores and information about which questions students missed. Holt Online Assessment is available through the Premier Online Edition of *Holt California Earth Science.*

Holt Anthology of Science Fiction

The Holt Anthology of Science Fiction includes thought-provoking stories that are relevant to science instruction. Enhance students' learning by asking them to read a story from the *Holt Anthology of Science Fiction* and to answer questions about what they have read.

16 **Pacing**
- **This chapter should take approximately 9.5 days to complete.**
- **Getting Started should take approximately 1 day to complete.**

Chapter 16 Interactions of Living Things

The Big Idea Organisms interact with each other and with the nonliving parts of their environment.

This chapter was designed to cover the California Grade 6 Science Standards about the ways organisms interact with each other and with their environment (6.5.a, 6.5.b, 6.5.c, and 6.5.e). The chapter describes how energy and matter are transferred from one organism to another and between organisms and their environment. It also describes how organisms can be categorized by the functions that the organisms serve in their ecosystems.

After they have completed this chapter, students will begin a chapter about biomes and ecosystems.

Getting Started

		Teach
	Standards Course of Study	❑ **SE Explore Activity** Who Eats Whom?, p. 549 ❑ **Datasheet B for Explore Activity** Who Eats Whom? ■
Universal Access	Advanced Learners/GATE	❑ **Chapter Starter Transparency** ❑ **Datasheet C for Explore Activity** Who Eats Whom? ■
Universal Access	Basic Learners	❑ **SE Improving Comprehension,** p. 546 ❑ **Chapter Starter Transparency** ❑ **Datasheet A for Explore Activity** Who Eats Whom? ■
Universal Access	English Learners	❑ **SE Improving Comprehension,** p. 546 ❑ **SE Unpacking the Standards,** p. 547
Universal Access	Special Education Students	❑ **SE Improving Comprehension,** p. 546 ❑ **SE Unpacking the Standards,** p. 547
Universal Access	Struggling Readers	❑ **SE Improving Comprehension,** p. 546

Key

SE Student Edition
TE Teacher's Edition

Chapter Resource File
Workbook

CD or CD-ROM
Transparency

Video
■ Also available in Spanish

All resources listed below are also available on the One-Stop Planner.

The California Science Standards listed below are covered in this chapter:

Focus on Earth Sciences

6.5.a Students know energy entering ecosystems as sunlight is transferred by producers into chemical energy through photosynthesis and then from organism to organism through food webs.

6.5.b Students know matter is transferred over time from one organism to others in the food web and between organisms and the physical environment.

6.5.c Students know populations of organisms can be categorized by the functions they serve in an ecosystem.

6.5.e Students know the number and types of organisms an ecosystem can support depends on the resources available and on abiotic factors, such as quantities of light and water, a range of temperatures, and soil composition.

Investigation and Experimentation

6.7.c Construct appropriate graphs from data and develop qualitative statements about the relationships between variables.

6.7.e Recognize whether evidence is consistent with a proposed explanation.

Practice

- ❑ SE **Organize Activity** Tri-Fold, p. 548
- ❑ TE **Word Parts,** p. 547
- ❑ TE **Using Other Graphic Organizers,** p. 546

Assess

- ❑ **Chapter Pretest**

Pacing
• This section should take approximately 1.5 days to complete.

Section 1 Everything Is Connected

Key Concept Organisms depend on each other and on the nonliving resources in their environment.

	Teach
Standards Course of Study	❑ **Bellringer Transparency** ❑ **PowerPoint® Resources** ❑ **E81** The Six Levels of Environmental Organization ❑ **E82** A Salt Marsh Ecosystem ❑ **SE Quick Lab** Meeting the Neighbors, p. 551 ❑ **Datasheet B for Quick Lab** Meeting the Neighbors ■
Universal Access: Advanced Learners/GATE	❑ **Datasheet C for Quick Lab** Meeting the Neighbors ■
Universal Access: Basic Learners	❑ **Datasheet A for Quick Lab** Meeting the Neighbors ■
Universal Access: English Learners	
Universal Access: Special Education Students	
Universal Access: Struggling Readers	❑ **SE Reading Strategy** Outlining, p. 550

Key

SE Student Edition
TE Teacher's Edition
Chapter Resource File
Workbook
CD or CD-ROM
Transparency
Video
■ Also available in Spanish

All resources listed below are also available on the One-Stop Planner.

Focus on Earth Sciences: 6.5.b, 6.5.e
Math: Algebra and Functions 6.2.1
English–Language Arts: Reading 6.2.4

Practice	Assess
❑ SE **Section Review,** p. 553 ❑ **Section Review** ■	❑ SE **Standards Checks,** pp. 550, 552 ❑ TE **Standards Focus,** p. 552 • Assess • Reteach • Re-Assess ❑ **Section Quiz** ■
❑ TE **Wordwise** Word Parts, p. 551 ❑ TE **Using the Figure** Levels of Organization, p. 552 ❑ **SciLinks Activity**	❑ **Long-Term Projects & Research Ideas** Out of House and Home
❑ TE **Group Activity** Exploring School Grounds, p. 550 ❑ TE **Using the Figure** Levels of Organization, p. 552 ❑ TE **Universal Access** Nature Scene, p. 553 ❑ **Interactive Reader and Study Guide** ❑ **Vocabulary and Section Summary A** ■	
❑ TE **Group Activity** Exploring School Grounds, p. 550 ❑ TE **Wordwise Word Parts,** p. 551 ❑ TE **Using the Figure** Levels of Organization, p. 552 ❑ **Interactive Reader and Study Guide** ❑ **Vocabulary and Section Summary A** ■ ❑ **Vocabulary and Section Summary B**	
❑ **Interactive Reader and Study Guide**	❑ TE **Universal Access** Salt Marsh Search, p. 552
❑ TE **Universal Access** Converting Text to Diagrams, p. 550 ❑ TE **Wordwise** Word Parts, p. 551 ❑ TE **Using the Figure** Levels of Organization, p. 552 ❑ **Interactive Reader and Study Guide** ❑ **Directed Reading A** ■ ❑ **Directed Reading B**	

Pacing • This section should take approximately 1.5 days to complete.

Section 2 Living Things Need Energy

Key Concept Energy and matter flow between organisms and their environment.

	Teach
Standards Course of Study	❑ **Bellringer Transparency** ❑ **PowerPoint® Resources** ❑ **E83** A Food Chain ❑ **E84** A Food Web ❑ **E85** Energy Pyramid ❑ SE **Quick Lab,** How Are the Organisms in a Food Chain Connected?, p. 558 ❑ **Datasheet B for Quick Lab** How Are the Organisms in a Food Chain Connected? ■
Universal Access — Advanced Learners/GATE	❑ **L17 Link to Life Science** The Connection Between Photosynthesis and Cellular Respiration ❑ TE **Discussion** Learning Vocabulary, p. 555 ❑ **Datasheet C for Quick** How Are the Organisms in a Food Chain Connected? ■
Basic Learners	❑ **Datasheet A for Quick Lab** How Are the Organisms in a Food Chain Connected? ■ ❑ **EcoLabs & Field Activities** Ditch's Brew
English Learners	❑ TE **Discussion** Learning Vocabulary, p. 555
Special Education Students	
Struggling Readers	❑ SE **Reading Strategy** Graphic Organizer, p. 554 ❑ TE **Discussion** Learning Vocabulary, p. 555

Key

SE Student Edition
TE Teacher's Edition
Chapter Resource File
Workbook
CD or CD-ROM
Transparency
Video
■ Also available in Spanish

All resources listed below are also available on the One-Stop Planner.

Focus on Earth Sciences: 6.5.a, 6.5.b, 6.5.c, 6.5.e
Math: Number Sense 6.1.4
English–Language Arts: Reading 6.2.4

Practice	Assess
❑ SE **Section Review,** p. 559 ❑ **Section Review** ■	❑ SE **Standards Checks,** pp. 554, 556, 557, 558 ❑ TE **Standards Focus,** p. 558 • Assess • Reteach • Re-Assess ❑ **Section Quiz** ■
❑ TE **Connection to Math** Energy Loss, p. 556 ❑ TE **Universal Access** Shopping List Sources, p. 557 ❑ TE **Activity** Inferring Information, p. 557	❑ TE **Using the Figure** Energy Transfer, p. 554 ❑ TE **Connection to Real World** Cockroaches, p. 555 ❑ TE **Homework** Energy Walk, p. 555 ❑ TE **Research** Peregrine Falcons, p. 557
❑ TE **Activity** Food Sources, p. 554 ❑ **Interactive Reader and Study Guide** ❑ **Vocabulary and Section Summary A** ■	❑ TE **Using the Figure** Energy Transfer, p. 554 ❑ TE **Homework** Energy Walk, p. 555
❑ TE **Universal Access** Oral Communication, p. 556 ❑ **Interactive Reader and Study Guide** ❑ **Vocabulary and Section Summary A** ■ ❑ **Vocabulary and Section Summary B** ❑ **EcoLabs & Field Activities** Survival is Just a Roll of the Dice	❑ TE **Using the Figure** Energy Transfer, p. 554 ❑ TE **Connection to Real World** Cockroaches, p. 555 ❑ TE **Homework** Energy Walk, p. 555
❑ TE **Activity** Food Sources, p. 554 ❑ TE **Universal Access** Who Eats Whom?, p. 556 ❑ **Interactive Reader and Study Guide**	
❑ TE **Universal Access** Identifying Structural Patterns, p. 555 ❑ **Interactive Reader and Study Guide** ❑ **Directed Reading A** ■ ❑ **Directed Reading B**	❑ TE **Homework** Energy Walk, p. 555

 • This section should take approximately 1.5 days to complete.

Section 3 Types of Interactions

Key Concept Organisms depend on their relationships with each other and on the resources in their environment.

	Teach
Standards Course of Study	❑ **Bellringer Transparency** ❑ **PowerPoint® Resources** ❑ SE **Quick Lab** Predator or Prey?, p. 563 ❑ **Datasheet B for Quick Lab** Predator or Prey? ■
Universal Access: Advanced Learners/GATE	❑ TE **Discussion** Exploring Relationships, p. 560 ❑ TE **Discussion** Yawning, p. 563 ❑ **Datasheet C for Quick Lab** Predator or Prey? ■
Universal Access: Basic Learners	❑ TE **Universal Access** Adaptations, p. 560 ❑ TE **Discussion** Yawning, p. 563 ❑ **Datasheet A for Quick Lab** Predator or Prey? ■
Universal Access: English Learners	❑ TE **Discussion** Exploring Relationships, p. 560 ❑ TE **Discussion** Yawning, p. 563
Universal Access: Special Education Students	❑ TE **Discussion** Yawning, p. 563
Universal Access: Struggling Readers	❑ SE **Reading Strategy** Summarizing, p. 560

Key

SE Student Edition
TE Teacher's Edition

Chapter Resource File
Workbook

CD or CD-ROM
Transparency

Video
■ Also available in Spanish

All resources listed below are also available on the One-Stop Planner.

Focus on Earth Sciences: 6.5.b, 6.5.e
English–Language Arts: Reading 6.2.4

Practice	Assess
❑ SE **Section Review,** p. 565 ❑ **Section Review** ■	❑ SE **Standards Checks,** pp. 560, 561, 565 ❑ TE **Standards Focus,** p. 564 • Assess • Reteach • Re-Assess ❑ **Section Quiz** ■
❑ TE **Universal Access** Deer Debate, p. 561 ❑ TE **Activity** Predators and Prey, p. 562 ❑ **Critical Thinking**	❑ TE **Homework** Diagramming a Picture, p. 561
❑ TE **Activity** Predators and Prey, p. 562 ❑ TE **Using the Figure** Camouflage, p. 562 ❑ **Interactive Reader and Study Guide** ❑ **Vocabulary and Section Summary A** ■ ❑ **Reinforcement Worksheet**	❑ TE **Homework** Diagramming a Picture, p. 561
❑ TE **Universal Access** Vocabulary Focus, p. 561 ❑ TE **Activity** Predators and Prey, p. 562 ❑ TE **Using the Figure** Camouflage, p. 562 ❑ **Interactive Reader and Study Guide** ❑ **Vocabulary and Section Summary A** ■ ❑ **Vocabulary and Section Summary B**	
❑ TE **Using the Figure** Camouflage, p. 562 ❑ TE **Universal Access** Predator/Prey Chains, p. 562 ❑ **Interactive Reader and Study Guide**	❑ TE **Homework** Diagramming a Picture, p. 561
❑ TE **Using the Figure** Camouflage, p. 562 ❑ TE **Universal Access** Identifying Main Ideas, p. 563 ❑ **Interactive Reader and Study Guide** ❑ **Directed Reading A** ■ ❑ **Directed Reading B**	

Pacing

- Chapter Lab, Review, and Assessment should take approximately 4 days to complete.

Wrapping Up

		Teach
	Standards Course of Study	❑ SE **Skills Practice Lab** Too Much of a Nutrient?, pp. 566–567 ❑ **Datasheet B for Chapter Lab** Too Much of a Nutrient? ■ ❑ **Standards Review Transparency** ■
Universal Access	Advanced Learners/GATE	❑ TE **Teaching Strategy,** p. 575 ❑ **Datasheet C for Chapter Lab** Too Much of a Nutrient? ■
	Basic Learners	❑ **Datasheet A for Chapter Lab** Too Much of a Nutrient? ■
	English Learners	❑ TE **Teaching Strategy,** p. 575
	Special Education Students	
	Struggling Readers	

Additional Resources

SUPER SUMMARY

Have students review the major concepts in this chapter by using the Super Summary that includes the following:

- an outline of important points in the chapter
- flashcards for chapter vocabulary
- an interactive quiz

Go to **go.hrw.com**
Type in the keyword HY7INTS

Performance-Based Assessments

The Chapter Resource File for this chapter contains a hands-on activity that can be used to help assess student progress in a nontraditional format. In the Performance-Based Assessment for this chapter, students discuss the habitats of different animals.

Key

- **SE** Student Edition
- **TE** Teacher's Edition
- Chapter Resource File
- Workbook
- CD or CD-ROM
- Transparency
- Video
- ■ Also available in Spanish

All resources listed below are also available on the One-Stop Planner.

Focus on Earth Sciences: 6.5.b, 6.5.e, 6.7.c, 6.7.e
Math: Number Sense 6.1.4
English–Language Arts: Writing 6.1.3

Practice	Assess
❑ **SE** **Science Skills Activity** Constructing a Bar Graph, p. 568 ❑ **Datasheet for Science Skills Activity** ■ ❑ **Concept Mapping Transparency** ❑ **SE** **Chapter Review,** pp. 570–571 ❑ **Chapter Review** ■	❑ **SE** **Standards Assessment,** pp. 572–573 ❑ **Standards Assessment** ❑ **Chapter Test B** ■ ❑ **Standards Review Workbook** ■
❑ **TE** **Identifying Prefixes and Suffixes** Vocabulary Affixes, p. 569	❑ **SE** **Math Activity,** p. 574 ❑ **SE** **Social Studies Activity,** p. 575 ❑ **TE** **Focus on Writing** Short Story, p. 569 ❑ **Chapter Test C** ❑ **Brain Food Video Quiz**
❑ **SE** **Language Arts Activity,** p. 574 ❑ **TE** **Activity** Dog Discussion, p. 574	❑ **TE** **Focus on Writing** Short Story, p. 569 ❑ **Chapter Test A** ❑ **Brain Food Video Quiz**
❑ **SE** **Language Arts Activity,** p. 574 ❑ **TE** **Identifying Prefixes and Suffixes** Vocabulary Affixes, p. 569 ❑ **TE** **Activity** Dog Discussion, p. 574	❑ **SE** **Math Activity,** p. 574 ❑ **TE** **Focus on Writing** Short Story, p. 569 ❑ **Brain Food Video Quiz**
❑ **TE** **Universal Access** Grid Adaptation, p. 568	❑ **Brain Food Video Quiz**
❑ **TE** **Universal Access** Reading Tables, p. 568 ❑ **TE** **Identifying Prefixes and Suffixes** Vocabulary Affixes, p. 569	❑ **SE** **Math Activity,** p. 574 ❑ **Brain Food Video Quiz**

Holt Online Assessment

Post tests and quizzes to Holt Online Assessment, an assessment management tool. The system automatically grades the assessments, and you receive students' scores and information about which questions students missed. Holt Online Assessment is available through the Premier Online Edition of *Holt California Earth Science.*

Holt Anthology of Science Fiction

The Holt Anthology of Science Fiction includes thought-provoking stories that are relevant to science instruction. Enhance students' learning by asking them to read a story from the *Holt Anthology of Science Fiction* and to answer questions about what they have read.

Pacing

- This chapter should take approximately 11 days to complete.
- Getting Started should take approximately 1 day to complete.

Chapter 17 Biomes and Ecosystems

The Big Idea Biomes and ecosystems are characterized by their living and nonliving parts.

This chapter was designed to cover the California Grade 6 Science Standards about biomes and ecosystems (6.5.a, 6.5.b, 6.5.c, 6.5.d, and 6.5.e). It follows the chapter about the interactions of living things, because understanding how organisms depend on each other and their environment provides a base on which students can build an understanding of how biomes and ecosystems function. The study of ecology will help prepare students for the study of life sciences in grade 7.

Getting Started

		Teach
	Standards Course of Study	❑ SE **Explore Activity** Build a Mini-Ecosystem, p. 579 ❑ **Datasheet B for Explore Activity** Build a Mini-Ecosystem ■
Universal Access	Advanced Learners/GATE	❑ **Chapter Starter Transparency** ❑ **Datasheet C for Explore Activity** Build a Mini-Ecosystem ■
Universal Access	Basic Learners	❑ SE **Improving Comprehension,** p. 576 ❑ **Chapter Starter Transparency** ❑ **Datasheet A for Explore Activity** Build a Mini-Ecosystem ■
Universal Access	English Learners	❑ SE **Improving Comprehension,** p. 576 ❑ SE **Unpacking the Standards,** p. 577
Universal Access	Special Education Students	❑ SE **Improving Comprehension,** p. 576 ❑ SE **Unpacking the Standards,** p. 577
Universal Access	Struggling Readers	❑ SE **Improving Comprehension,** p. 576

Key

SE Student Edition
TE Teacher's Edition
Chapter Resource File
Workbook
CD or CD-ROM
Transparency
Video
■ Also available in Spanish

All resources listed below are also available on the One-Stop Planner.

The California Science Standards listed below are covered in this chapter:

Focus on Earth Sciences

6.5.a Students know energy entering ecosystems as sunlight is transferred by producers into chemical energy through photosynthesis and then from organism to organism through food webs.

6.5.b Students know matter is transferred over time from one organism to others in the food web and between organisms and the physical environment.

6.5.c Students know populations of organisms can be categorized by the functions they serve in an ecosystem.

6.5.d Students know different kinds of organisms may play similar ecological roles in similar biomes.

6.5.e Students know the number and types of organisms an ecosystem can support depends on the resources available and on abiotic factors, such as quantities of light and water, a range of temperatures, and soil composition.

Investigation and Experimentation

6.7.a Develop a hypothesis.

6.7.b Select and use appropriate tools and technology (including calculators, computers, balances, spring scales, microscopes, and binoculars) to perform tests, collect data, and display data.

6.7.e Recognize whether evidence is consistent with a proposed explanation.

Practice	Assess
❑ SE **Organize Activity** Three-Panel Flip Chart, p. 578	❑ **Chapter Pretest**
❑ TE **Words With Multiple Meanings,** p. 577	
❑ TE **Using Other Graphic Organizers,** p. 576	

17

Pacing

• This section should take approximately 1 day to complete.

Section 1 Studying the Environment

Key Concept In every environment, organisms depend on each other and on the nonliving things in the environment to survive.

		Teach
	Standards Course of Study	❑ **Bellringer Transparency** ❑ **PowerPoint® Resources** ❑ **E86** Land Biomes ❑ SE **Quick Lab** Organisms and Water Resources, p. 582 ❑ **Datasheet B for Quick Lab** Organisms and Water Resources ■
Universal Access	Advanced Learners/GATE	❑ TE **Discussion** Abiotic Effects, p. 581 ❑ **Datasheet C for Quick Lab** Organisms and Water Resources ■
Universal Access	Basic Learners	❑ **Datasheet A for Quick Lab** Organisms and Water Resources ■
Universal Access	English Learners	❑ TE **Discussion** Abiotic Effects, p. 581
Universal Access	Special Education Students	
Universal Access	Struggling Readers	❑ SE **Reading Strategy** Graphic Organizer, p. 580 ❑ TE **Universal Access** Using Context Clues, p. 580

Additional Resources

Holt Lab Generator CD-ROM

Search for any lab by topic, standard, difficulty level, or time. Edit any lab to fit your needs, or create your own labs. Use the Lab Materials QuickList software to customize your lab materials list. Lab datasheets are also available in Spanish on this CD-ROM.

Guided Reading Audio CD Program

The Guided Reading Audio CD Program provides a direct reading of the student text. This resource is helpful to auditory learners and struggling readers. This program is available in English and Spanish.

Key

SE Student Edition
TE Teacher's Edition
Chapter Resource File
Workbook
CD or CD-ROM
Transparency
Video
■ Also available in Spanish

All resources listed below are also available on the One-Stop Planner.

Focus on Earth Sciences: 6.5.a, 6.5.b, 6.5.c, 6.5.d, 6.5.e
Math: Statistics, Data Analysis, and Probability 6.1.1
English–Language Arts: Reading 6.2.4

Practice	Assess
❑ SE **Section Review,** p. 583 ❑ **Section Review** ■	❑ SE **Standards Checks,** pp. 581, 582 ❑ TE **Standards Focus,** p. 582 • Assess • Reteach • Re-Assess ❑ **Section Quiz** ■
❑ TE **Using the Figure** Similar Places, p. 580 ❑ TE **Universal Access** Vocabulary Strategy, p. 580	
❑ **Interactive Reader and Study Guide** ❑ **Vocabulary and Section Summary A** ■	
❑ TE **Using the Figure** Similar Places, p. 580 ❑ TE **Universal Access** Writing About Organisms, p. 583 ❑ **Interactive Reader and Study Guide** ❑ **Vocabulary and Section Summary A** ■ ❑ **Vocabulary and Section Summary B**	
❑ TE **Universal Access** See, Hear, and Smell, p. 581 ❑ **Interactive Reader and Study Guide**	
❑ **Interactive Reader and Study Guide** ❑ **Directed Reading A** ■ ❑ **Directed Reading B**	

Reviewing Prior Knowledge

Prepare students to learn about how the environment is organized at the biome level and ecosystem level by brainstorming words that are associated with biomes. See page 580 of the Teacher's Edition.

Roles in a Biome Students may think that similar biomes around the world are filled with the same species of animals and plants. To correct this misconception, see page 581 in the Teacher's Edition.

Math Support

Science and math go hand in hand. The Math Skills item in the Section Review on page 583 helps students practice math skills in a scientific context.

Pacing • This section should take approximately 2 days to complete.

Section 2 Land Biomes

Key Concept The kinds of plants and animals that live in a biome are determined by the local climate.

		Teach
	Standards Course of Study	❑ **Bellringer Transparency** ❑ **PowerPoint® Resources** ❑ **E87** Desert ❑ **E88** Temperate Deciduous Forest ❑ **SE Quick Lab** What's Your Biome?, p. 590 ❑ **Datasheet B for Quick Lab** What's Your Biome? ■
Universal Access	Advanced Learners/GATE	❑ **TE Discussion** Native Grazers, p. 586 ❑ **Datasheet C for Quick Lab** What's Your Biome? ■
	Basic Learners	❑ **TE Demonstration** Chaparral Plants, p. 585 ❑ **TE Group Activity** Visual Precipitation, p. 586 ❑ **Datasheet A for Quick Lab** What's Your Biome? ■
	English Learners	❑ **TE Demonstration** Chaparral Plants, p. 585 ❑ **TE Discussion** Native Grazers, p. 586 ❑ **TE Group Activity** Visual Precipitation, p. 586
	Special Education Students	❑ **TE Demonstration** Chaparral Plants, p. 585 ❑ **TE Group Activity** Visual Precipitation, p. 586
	Struggling Readers	❑ **SE Reading Strategy** Graphic Organizer, p. 584 ❑ **TE Group Activity** Visual Precipitation, p. 586

Key

SE Student Edition
TE Teacher's Edition
Chapter Resource File
Workbook
CD or CD-ROM
Transparency
Video
■ Also available in Spanish

All resources listed below are also available on the One-Stop Planner.

Focus on Earth Sciences: 6.5.c, 6.5.d, 6.5.e
Math: Algebra and Functions 6.2.1
English–Language Arts: Reading 6.2.4

Practice	Assess
❑ SE **Section Review,** p. 591 ❑ **Section Review** ■	❑ SE **Standards Checks,** pp. 584, 585, 586, 587, 589, 590 ❑ TE **Standards Focus,** p. 590 • Assess • Reteach • Re-Assess ❑ **Section Quiz** ■
❑ **Long-Term Projects & Research Ideas** Tropical Medicine	❑ TE **Universal Access** Chaparral Fire Control, p. 585 ❑ TE **Universal Access** Melting Tundra, p. 587 ❑ TE **Homework** Taiga Report, p. 588 ❑ TE **Group Activity** Forest News, p. 589
❑ TE **Activity** The Biome Song, p. 587 ❑ TE **Group Activity** Biome Improvisation, p. 588 ❑ TE **Activity** Role Cards, p. 589 ❑ **Interactive Reader and Study Guide** ❑ **Vocabulary and Section Summary A** ■ ❑ **Reinforcement Worksheet**	❑ TE **Universal Access** Depicting Forests, p. 589
❑ TE **Activity** Describing Biomes, p. 584 ❑ TE **Activity** The Biome Song, p. 587 ❑ TE **Group Activity** Biome Improvisation, p. 588 ❑ TE **Activity** Role Cards, p. 589 ❑ **Interactive Reader and Study Guide** ❑ **Vocabulary and Section Summary A** ■ ❑ **Vocabulary and Section Summary B** ❑ **EcoLabs & Field Activities** Biome Adventure Travel	❑ TE **Homework** Taiga Report, p. 588 ❑ TE **Universal Access** Characteristics of Biomes, p. 589 ❑ TE **Group Activity** Forest News, p. 589
❑ TE **Universal Access** Grassland, Desert, and Tundra, p. 586 ❑ TE **Activity** The Biome Song, p. 587 ❑ **Interactive Reader and Study Guide**	❑ TE **Universal Access** Say, Define, and Use, p. 588
❑ TE **Universal Access** Clarifying Understanding, p. 584 ❑ **Interactive Reader and Study Guide** ❑ **Directed Reading A** ■ ❑ **Directed Reading B**	

17 Pacing

• This section should take approximately 2 days to complete.

Section 3 Marine Ecosystems

Key Concept Organisms in marine ecosystems depend on the abiotic factors and biotic factors in their environment.

		Teach
	Standards Course of Study	❑ **Bellringer Transparency** ❑ **PowerPoint® Resources** ❑ **E89** Ocean Temperature Zones ❑ **E90** Ocean Zones: A ❑ **E91** Ocean Zones: B ❑ **SE Quick Lab** How to Categorize Organisms, p. 598 ❑ **Datasheet B for Quick Lab** How to Categorize Organisms ■
Universal Access	**Advanced Learners/GATE**	❑ **L46 Link to Life Science** The Three Domains ❑ TE **Discussion** Oceans, p. 592 ❑ TE **Wordwise** Wandering Phytoplankton, p. 593 ❑ TE **Connection to Chemistry** Salinity, p. 596 ❑ **Datasheet C for Quick Lab** How to Categorize Organisms ■
Universal Access	**Basic Learners**	❑ **Datasheet A for Quick Lab** How to Categorize Organisms ■ ❑ TE **Using the Figure** Ocean Zones, p. 594
Universal Access	**English Learners**	❑ TE **Discussion** Oceans, p. 592 ❑ TE **Wordwise** Wandering Phytoplankton, p. 593 ❑ TE **Using the Figure** Ocean Zones, p. 594
Universal Access	**Special Education Students**	❑ TE **Using the Figure** Ocean Zones, p. 594
Universal Access	**Struggling Readers**	❑ SE **Reading Strategy** Outlining, p. 592 ❑ TE **Wordwise** Wandering Phytoplankton, p. 593

Key

SE Student Edition
TE Teacher's Edition
Chapter Resource File
Workbook
CD or CD-ROM
Transparency
Video
■ Also available in Spanish

All resources listed below are also available on the One-Stop Planner.

Focus on Earth Sciences: 6.5.a, 6.5.c, 6.5.e
Math: Number Sense 6.2.1; Algebra and Functions 6.2.1
English–Language Arts: Reading 6.2.4

Practice	Assess
❑ SE **Section Review,** p. 599 ❑ **Section Review** ■	❑ SE **Standards Checks,** pp. 592, 593, 595, 596, 597, 598 ❑ TE **Standards Focus,** p. 598 • Assess • Reteach • Re-Assess ❑ **Section Quiz** ■
❑ **SciLinks Activity** ❑ TE **Group Activity** Food Webs: A Continuing Story, p. 595 ❑ TE **Discussion** Similar Roles, p. 596	❑ TE **Homework** Intertidal Report, p. 596 ❑ TE **Universal Access** Hydrothermal Vent Report, p. 597
❑ TE **Using the Figure** Reading a Graph, p. 593 ❑ TE **Universal Access** Ocean Zone Interpretations, p. 595 ❑ TE **Group Activity** Food Webs: A Continuing Story, p. 595 ❑ **Interactive Reader and Study Guide** ❑ **Vocabulary and Section Summary A** ■	❑ TE **Cultural Awareness** Algal Food, p. 594 ❑ TE **Homework** Intertidal Report, p. 596
❑ TE **Using the Figure** Reading a Graph, p. 593 ❑ TE **Group Activity** Food Webs: A Continuing Story, p. 595 ❑ TE **Universal Access** Creating a Quiz Game: Ecosystems, p. 597 ❑ **Interactive Reader and Study Guide** ❑ **Vocabulary and Section Summary A** ■ ❑ **Vocabulary and Section Summary B**	❑ TE **Homework** Intertidal Report, p. 596
❑ TE **Universal Access** Show Me Where, p. 594 ❑ **Interactive Reader and Study Guide**	❑ TE **Universal Access** Animal Stories, p. 596
❑ TE **Using the Figure** Reading a Graph, p. 593 ❑ TE **Universal Access** Previewing, p. 594 ❑ **Interactive Reader and Study Guide** ❑ **Directed Reading A** ■ ❑ **Directed Reading B**	❑ TE **Homework** Intertidal Report, p. 596

 • This section should take approximately 1 day to complete.

Section 4 Freshwater Ecosystems

Key Concept Organisms in freshwater ecosystems depend on the abiotic and biotic factors in their environment.

	Teach
Standards Course of Study	❑ **Bellringer Transparency** ❑ **PowerPoint® Resources** ❑ **E92** Rivers ❑ **E93** Lake Zones ❑ **SE Quick Lab** Pond-Food Relationships, p. 601 ❑ **Datasheet B for Quick Lab** Pond-Food Relationships ■
Universal Access: Advanced Learners/GATE	❑ **TE Discussion,** p. 600 ❑ **Datasheet C for Quick Lab** Pond-Food Relationships ■
Universal Access: Basic Learners	❑ **Datasheet A for Quick Lab** Pond-Food Relationships ■
Universal Access: English Learners	❑ **TE Discussion,** p. 600
Universal Access: Special Education Students	
Universal Access: Struggling Readers	❑ **SE Reading Strategy** Graphic Organizer, p. 600

Additional Resources

Holt Lab Generator CD-ROM

Search for any lab by topic, standard, difficulty level, or time. Edit any lab to fit your needs, or create your own labs. Use the Lab Materials QuickList software to customize your lab materials list. Lab datasheets are also available in Spanish on this CD-ROM.

Guided Reading Audio CD Program

The Guided Reading Audio CD Program provides a direct reading of the student text. This resource is helpful to auditory learners and struggling readers. This program is available in English and Spanish.

Key

SE Student Edition
TE Teacher's Edition
Chapter Resource File
Workbook
CD or CD-ROM
Transparency
Video
■ Also available in Spanish

All resources listed below are also available on the One-Stop Planner.

Focus on Earth Sciences: 6.5.c, 6.5.e
Math: Number Sense 6.2.1
English–Language Arts: Reading 6.2.4

Practice	Assess
❑ SE **Section Review,** p. 603 ❑ **Section Review** ■	❑ SE **Standards Checks,** pp. 601, 603 ❑ TE **Standards Focus,** p. 602 • Assess • Reteach • Re-Assess ❑ **Section Quiz** ■
❑ TE **Universal Access** Pond Food Webs, p. 601 ❑ **Critical Thinking**	
❑ **Interactive Reader and Study Guide** ❑ **Vocabulary and Section Summary A** ■	❑ TE **Universal Access** Model Ecosystems, p. 600
❑ **Interactive Reader and Study Guide** ❑ **Vocabulary and Section Summary A** ■ ❑ **Vocabulary and Section Summary B**	
❑ **Interactive Reader and Study Guide**	
❑ TE **Universal Access** Understanding Word Families, p. 602 ❑ **Interactive Reader and Study Guide** ❑ **Directed Reading A** ■ ❑ **Directed Reading B**	

Reviewing Prior Knowledge

Prepare students to learn about how rivers form by using Figure 1. See page 600 of the Teacher's Edition.

MISCONCEPTION ALERT

Energy Flow Students may think that freshwater ecosystems are self-contained, and that energy cycles only within the system. To correct this misconception, see page 601 in the Teacher's Edition.

Math Support

Science and math go hand in hand. The Math Skills item in the Section Review on page 603 helps students practice math skills in a scientific context.

Pacing

- Chapter Lab, Review, and Assessment should take approximately 4 days to complete.

Wrapping Up

	Teach
Standards Course of Study	❑ SE **Inquiry Lab** Discovering Mini-Ecosystems, pp. 604–605 ❑ **Datasheet B for Chapter Lab** Discovering Mini-Ecosystems ■ ❑ **Standards Review Transparency** ■
Universal Access: Advanced Learners/GATE	❑ **Datasheet C for Chapter Lab** Discovering Mini-Ecosystems ■
Universal Access: Basic Learners	❑ **Datasheet A for Chapter Lab** Discovering Mini-Ecosystems ■
Universal Access: English Learners	
Universal Access: Special Education Students	
Universal Access: Struggling Readers	

Additional Resources

SUPER SUMMARY

Have students review the major concepts in this chapter by using the Super Summary that includes the following:

- an outline of important points in the chapter
- flashcards for chapter vocabulary
- an interactive quiz

Go to **go.hrw.com**
Type in the keyword HY7ECOS

Performance-Based Assessments

The Chapter Resource File for this chapter contains a hands-on activity that can be used to help assess student progress in a nontraditional format. In the Performance-Based Assessment for this chapter, students describe a location and choose items to survive there for 1 week.

Key

SE Student Edition
TE Teacher's Edition
Chapter Resource File
Workbook
CD or CD-ROM
Transparency
Video
■ Also available in Spanish

All resources listed below are also available on the One-Stop Planner.

Focus on Earth Sciences: 6.5.b, 6.5.e, 6.7.e
Math: Number Sense 6.2.1
English–Language Arts: Writing 6.1.3

Practice	Assess
❑ SE **Science Skills Activity** Organizing and Analyzing Evidence, p. 606 ❑ **Datasheet for Science Skills Activity** ■ ❑ **Concept Mapping Transparency** ❑ SE **Chapter Review,** pp. 608–609 ❑ **Chapter Review** ■	❑ SE **Standards Assessment,** pp. 610–611 ❑ **Standards Assessment** ❑ **Chapter Test B** ■ ❑ **Standards Review Workbook** ■
❑ TE **Identifying Roots** Word Find, p. 607 ❑ TE **Discussion,** p. 613	❑ TE **Focus on Writing** Rain Forest Story, p. 607 ❑ TE **Language Arts Activity,** p. 612 ❑ TE **Math Activity,** p. 612 ❑ TE **Social Studies Activity,** p. 613 ❑ **Chapter Test C** ❑ **Brain Food Video Quiz**
❑ TE **Identifying Roots** Word Find, p. 607	❑ TE **Focus on Writing** Rain Forest Story, p. 607 ❑ **Chapter Test A** ❑ **Brain Food Video Quiz**
❑ TE **Identifying Roots** Word Find, p. 607 ❑ TE **Discussion,** p. 613	❑ TE **Focus on Writing** Rain Forest Story, p. 607 ❑ TE **Language Arts Activity,** p. 612 ❑ TE **Social Studies Activity,** p. 613 ❑ **Brain Food Video Quiz**
❑ TE **Universal Access** Using a Computer, p. 606	❑ **Brain Food Video Quiz**
❑ TE **Universal Access** Converting Text to Table, p. 606 ❑ TE **Identifying Roots** Word Find, p. 607	❑ **Brain Food Video Quiz**

Holt Online Assessment

Post tests and quizzes to Holt Online Assessment, an assessment management tool. The system automatically grades the assessments, and you receive students' scores and information about which questions students missed. Holt Online Assessment is available through the Premier Online Edition of *Holt California Earth Science.*

Holt Anthology of Science Fiction

The Holt Anthology of Science Fiction includes thought-provoking stories that are relevant to science instruction. Enhance students' learning by asking them to read a story from the *Holt Anthology of Science Fiction* and to answer questions about what they have read.

Unpacking the Standards

The information below "unpacks" the standards by breaking them down into more basic parts. "What It Means" restates the standards as simply as possible."Why It Matters" explains why studying each standard is important.

California Standard	What It Means	Why It Matters
Plate Tectonics and Earth's Structure **6.1 Plate tectonics accounts for important features of Earth's surface and major geologic events.**	The movement of large pieces of Earth's surface shapes the surface of Earth and causes large geologic events, such as earthquakes and volcanoes.	Plate tectonics is a cornerstone of Earth science. This theory provides a basis for understanding many other Earth science topics.
6.1.a Students know evidence of plate tectonics is derived from the fit of the continents; the location of earthquakes, volcanoes, and midocean ridges; and the distribution of fossils, rock types, and ancient climatic zones.	Support for the idea that Earth's surface is made of slabs of rock that move around comes from how continents fit together, where earthquakes happen, where volcanoes and mid-ocean ridges are located, where rocks and fossils are found, and where climate was different in the geologic past.	To understand the theory of plate tectonics, students must understand how the theory developed and must understand that the theory is based on numerous types of convincing, reliable information.
6.1.b Students know Earth is composed of several layers: a cold, brittle lithosphere; a hot, convecting mantle; and a dense, metallic core.	Earth is made of several layers: a cold, brittle outer layer called the *lithosphere*; a hot, middle layer called the *mantle*, which circulates heat; and a dense core made of metals such as iron and nickel.	Scientists can use the information about the layers of Earth to explain how events, such as earthquakes and volcanoes, occur on the surface of Earth. They can also use this information to learn how Earth formed.
6.1.c Students know lithospheric plates the size of continents and oceans move at rates of centimeters per year in response to movements in the mantle.	Earth's crust and upper mantle are broken into large plates that are the size of oceans and continents. These plates move at rates of centimeters per year as the hot, convecting mantle moves.	By knowing the rate at which tectonic plates move, scientists can determine the rate at which surface features, such as mountains, are forming. Scientists also can predict future changes on Earth's surface and can predict approximately when those features will form.
6.1.d Students know that earthquakes are sudden motions along breaks in the crust called faults and that volcanoes and fissures are locations where magma reaches the surface.	Earthquakes are sudden movements along breaks in Earth's surface called *faults*, and volcanoes are places where molten rock reaches the surface.	Knowing where and why earthquakes and volcanoes happen helps people make wise decisions about how to live safely in fault zones and volcanically active areas.
6.1.e Students know major geologic events, such as earthquakes, volcanic eruptions, and mountain building, result from plate motions.	Important global events, such as earthquakes, volcanic eruptions, and mountain building, are caused by the movement of Earth's crust and upper mantle.	The movement of tectonic plates changes the shape of Earth's surface and can cause destruction by earthquakes and volcanic eruptions.

California Standard	What It Means	Why It Matters
6.1.f Students know how to explain major features of California geology (including mountains, faults, volcanoes) in terms of plate tectonics.	You must know how to explain how the movement of tectonic plates affected the formation of rocks and landforms (such as mountains, faults, and volcanoes) in California.	Plate tectonics has been the most important factor in the formation of landforms in California. These processes continue to shape the landscape of California.
6.1.g Students know how to determine the epicenter of an earthquake and know that the effects of an earthquake on any region vary, depending on the size of the earthquake, the distance of the region from the epicenter, the local geology, and the type of construction in the region.	You must know how to find an earthquake's starting point and must know that how much damage an earthquake causes depends on the size of the earthquake, the distance from the starting point, the type of rock under a place, the materials used in buildings, and the way in which the buildings were made.	Knowing where the epicenter of an earthquake is located helps scientists predict the future locations of earthquakes. Understanding factors that affect destruction caused by an earthquake helps scientists and officials decrease potential destruction caused by future earthquakes.
Shaping Earth's Surface **6.2 Topography is reshaped by the weathering of rock and soil and by the transportation and deposition of sediment.**	The shape of the land surface is changed by the breaking down of rock and soil and by the movement of rock and soil across the surface of the land.	Many features of Earth's landscape are shaped by the processes of weathering, transporting, and depositing sediment. Sometimes, these processes can be dramatic and can cause destruction and death.
6.2.a Students know water running downhill is the dominant process in shaping the landscape, including California's landscape.	Water running in rivers and over the land is the most important force in changing the natural shape of Earth's surface.	By learning how water has shaped the surface of Earth in the past, people can understand how water continues to shape Earth's surface.
6.2.b Students know rivers and streams are dynamic systems that erode, transport sediment, change course, and flood their banks in natural and recurring patterns.	Rivers and streams wear away and move rock and soil fragments, change their paths, and overflow their banks in natural patterns that happen over and over and year after year.	Houses built near streams or rivers are at risk for damage caused by flooding or erosion of stream banks. This danger can be sudden and unpredictable.
6.2.c Students know beaches are dynamic systems in which the sand is supplied by rivers and moved along the coast by the action of waves.	Beaches are always changing. Rivers carry sand to beaches and oceans, and ocean waves move the sand along the shoreline.	People who build houses or buildings on or near a beach or on the cliffs above a beach are in danger from ocean waves, which can move sand along the beach or undercut the cliffs to cause landslides.
6.2.d Students know earthquakes, volcanic eruptions, landslides, and floods change human and wildlife habitats.	Earthquakes, volcanic eruptions, landslides, and floods change the surroundings in which humans and wildlife live.	People must understand how and why volcanoes, landslides, and floods can change the land so that they can plan ahead for these natural events.
Heat (Thermal Energy) (Physical Sciences) **6.3 Heat moves in a predictable flow from warmer objects to cooler objects until all the objects are at the same temperature.**	Thermal energy (heat) always moves from a warmer item to a cooler item until both items have the same average kinetic energy (temperature).	Understanding how heat is transferred can help people control the temperature of things, such as their house, their food, and their body.
6.3.a Students know energy can be carried from one place to another by heat flow or by waves, including water, light and sound waves, or by moving objects.	Energy is moved from one place to another by heat flow, by waves, or by the movement of objects. Water waves, light waves, and sound waves are kinds of waves.	Understanding how energy is transferred helps students understand how the energy from a tsunami causes destruction, how ultraviolet light waves cause sunburn, and how sound waves bring music to their ears.

California Standard	What It Means	Why It Matters
6.3.b Students know that when fuel is consumed, most of the energy released becomes heat energy.	When fuel, such as wood, is burned, most of the energy that is given off is heat.	The heat released from the burning of fuel can be used to heat homes, generate electricity, and power cars.
6.3.c Students know heat flows in solids by conduction (which involves no flow of matter) and in fluids by conduction and by convection (which involves flow of matter).	When heat flows in solids, the heat can be transferred without the movement of matter. When heat flows in liquids and gases, the heat can be transferred with or without the movement of matter.	Convection and conduction are two processes for transferring heat. Without convection and conduction, people could not do many things, such as cook food or warm the air in their homes.
6.3.d Students know heat energy is also transferred between objects by radiation (radiation can travel through space).	Heat and energy can move from one place to another by radiation, the process by which energy moves through matter or through empty space.	Radiation is the process by which Earth is warmed by the sun. It is also the process by which a heater transfers heat to the air in a home.
Energy in the Earth System **6.4 Many phenomena on Earth's surface are affected by the transfer of energy through radiation and convection currents.**	Many processes on Earth's surface are driven by the movement of heat and energy by radiation and convection.	The transfer of energy causes the water cycle, changes in weather, ocean currents, and the movement of tectonic plates.
6.4.a Students know the sun is the major source of energy for phenomena on Earth's surface; it powers winds, ocean currents, and the water cycle.	The sun is the main source of energy for processes on Earth. Energy from the sun powers winds, ocean currents, and the water cycle.	Understanding that the sun is the main source of energy on Earth allows students to understand how changes in the distribution of solar energy on Earth could change processes on Earth.
6.4.b Students know solar energy reaches Earth through radiation, mostly in the form of visible light.	Energy from the sun reaches Earth through radiation. Most of this energy is light that humans can see.	There is no matter between Earth and the sun. Therefore, energy from the sun can reach Earth only by the process of radiation. Solar radiation is important to many processes that occur on Earth.
6.4.c Students know heat from Earth's interior reaches the surface primarily through convection.	Heat from deep inside Earth reaches Earth's surface mainly by the movement of hot material in Earth's mantle.	Convection currents inside Earth cause tectonic plate movements. These movements cause earthquakes, volcanic eruptions, and mountain building.
6.4.d Students know convection currents distribute heat in the atmosphere and oceans.	The movement of air in the atmosphere and of water in the ocean carries heat throughout the atmosphere and oceans.	The distribution of heat in the oceans and the atmosphere affects weather patterns around the world.
6.4.e Students know differences in pressure, heat, air movement, and humidity result in changes of weather.	Changes in the weight of the atmosphere, heat, air movement, and the amount of water in the air cause the conditions of the atmosphere to change.	By measuring changes in pressure, heat, air movement, and humidity, meteorologists can predict weather, which can save lives.
Ecology (Life Sciences) **6.5 Organisms in ecosystems exchange energy and nutrients among themselves and with the environment.**	Living things transfer energy and nutrients with each other and with their surroundings.	Humans are part of the environment and exchange energy with organisms around them and with their environment.
6.5.a Students know energy entering ecosystems as sunlight is transferred by producers into chemical energy through photosynthesis and then from organism to organism through food webs.	Producers, such as plants, change energy from the sun into chemical energy through photosynthesis. Then, this chemical energy passes to and between organisms that eat other organisms.	The sun is the ultimate source of energy for all living things on Earth. Humans are part of the food webs that transfer this energy.

California Standard	What It Means	Why It Matters
6.5.b Students know matter is transferred over time from one organism to others in the food web and between organisms and the physical environment.	Matter moves from one living thing to another in a food web. Matter also moves between living things and their nonliving surroundings.	Nutrients are forms of matter that are needed for organisms to survive. If the cycling of these nutrients in the food web and physical environment did not occur, it would be difficult for life to exist.
6.5.c Students know populations of organisms can be categorized by the functions they serve in an ecosystem.	Living things can be identified by the role that they play, or what they do, in an ecosystem.	The functions that an organism performs in an ecosystem is important for the survival of other organisms in the ecosystem.
6.5.d Students know different kinds of organisms may play similar ecological roles in similar biomes.	Different kinds of living things may have the same function in different environments that have similar characteristics.	By understanding the roles that organisms play in their environment, you can compare organisms in different environments.
6.5.e Students know the number and types of organisms an ecosystem can support depends on the resources available and on abiotic factors, such as quantities of light and water, a range of temperatures, and soil composition.	The number and kinds of living things that can live in an ecosystem depend on the available resources and on nonliving things, such as the amount of light and water, the high and low temperatures, and the makeup of the soil.	The world contains a limited number of resources. Therefore, Earth can support only a certain number of living things, including humans.
Resources **6.6 Sources of energy and materials differ in amounts, distribution, usefulness, and the time required for their formation.**	Resources used to generate energy and materials differ in how much of each resource is available, where each resource is found, whether it is useful, and how long it takes to form.	The location, amount, and use of energy resources, such as oil, affect the environment, the world economy, and political relations between countries.
6.6.a Students know the utility of energy sources is determined by factors that are involved in converting these sources to useful forms and the consequences of the conversion process.	The usefulness of energy resources is controlled by how these resources are changed into useful forms and what happens to the resources and the environment after the resources are changed.	Energy resources, such as oil, are used to power cars, heat homes, provide electricity, and make products. However, converting oil into these useful forms may damage the environment.
6.6.b Students know different natural energy and material resources, including air, soil, rocks, minerals, petroleum, fresh water, wildlife, and forests, and know how to classify them as renewable or nonrenewable.	You must be able to identify different natural resources, such as air, soil, rocks, minerals, petroleum, fresh water, wildlife, and forests, and be able to identify whether each resource can be replaced rapidly or not.	If humans are not careful about the consumption of natural resources, these resources will be unavailable in the future. Without natural resources, such as fresh air and water, humans would not be able to survive.
6.6.c Students know the natural origin of the materials used to make common objects.	You must be able to identify the natural resources that are used to make common objects.	When consuming natural resources, people must realize where the resources came from and how they were taken out of the environment.

California Standard	What It Means	Why It Matters
Investigation and Experimentation **6.7 Scientific progress is made by asking meaningful questions and conducting careful investigations.**	To advance science and technology, scientists must ask important questions and carry out detailed, well-planned research projects or experiments.	Understanding that scientific investigations have firm rules and standards allows people to trust the results of scientific investigations.
6.7.a Develop a hypothesis.	Write a possible explanation for or answer to a scientific research question that can be tested.	Scientists use hypotheses as a basis for planning and conducting experiments and for making observations.
6.7.b Select and use appropriate tools and technology (including calculators, computers, balances, spring scales, microscopes, and binoculars) to perform tests, collect data, and display data.	Choose the correct tools and technology (including calculators, computers, balances, spring scales, microscopes, and binoculars) to perform an experiment. Use these tools and technology to collect facts and figures, and show your research findings.	Scientific research depends on accurate and precise collection of facts and figures (data). Scientists must choose the correct tools and technology to get reliable, accurate results.
6.7.c Construct appropriate graphs from data and develop qualitative statements about the relationships between variables.	Make the correct type of graph to show your facts and figures. Then, write a statement that explains how one variable changes relative to the other variable.	To accurately show data, one must be able to choose the graph that most clearly and accurately shows the results. One must be able to read the graph to explain what the graph shows.
6.7.d Communicate the steps and results from an investigation in written reports and oral presentations.	Clearly explain the steps and the results of an experiment by using written reports and oral presentations.	Scientists share the results of their research with other scientists. To be effective, scientists must be able to write clearly and to speak to people effectively.
6.7.e Recognize whether evidence is consistent with a proposed explanation.	Figure out if observations and information agree or disagree with your previous ideas or explanations.	If evidence supports an explanation, the explanation may be true. If the evidence does not support the explanation, the hypothesis may need to be revised.
6.7.f Read a topographic map and a geologic map for evidence provided on the maps and construct and interpret a simple scale map.	Read maps that tell you what the shape of the land is and where rock units and structures are located. Make and read a simple map that is drawn to the correct proportions.	Reading maps is an important skill. People use maps to find locations, to navigate from one place to another, and to display information about Earth.
6.7.g Interpret events by sequence and time from natural phenomena (e.g., the relative ages of rocks and intrusions).	Explain the order of events and time by using natural processes and events. For example, you must be able to tell the ages of rocks and intrusions by the order in which these features formed.	Scientists use the order of events to figure out how Earth's processes work and to learn about Earth's history.
6.7.h Identify changes in natural phenomena over time without manipulating the phenomena (e.g., a tree limb, a grove of trees, a stream, a hillslope).	Point out the changes in natural processes or events over time without directly causing changes in the processes or events. For example, you must be able to see how a single tree, a group of trees, a stream, or the slope of a hill changes over time.	Scientists make many observations that allow them to determine how Earth's processes work. Observing the world around us also helps us make wise decisions and avoid danger.

Teaching Notes

Teaching Notes

Teaching Notes

Teaching Notes